MATERIALS FOR ENGINEERS

This text is intended for a first undergraduate course in materials science and engineering with an emphasis on mechanical and electrical properties. The text features numerous useful examples and exercises. It differs from other available texts in that it covers topics of greatest interest in most undergraduate programs, leaving more specialized and advanced coverage for later course books. The text begins with phases and phase diagrams. This is followed by a chapter on diffusion, which treats diffusion in multiphase as well as single-phase systems. The next several chapters on mechanical behavior and failure should be of particular interest to mechanical engineers. There are chapters on iron and steel and on nonferrous alloys, followed by chapters on specific types of materials. There is an emphasis on manufacturing, including recycling, casting and welding, powder processing, solid forming, and more modern techniques, including photolithography, vapor deposition, and the use of lasers.

William F. Hosford is a professor emeritus of materials science and engineering at the University of Michigan. Professor Hosford is the author of a number of books, including the leading selling *Metal Forming: Mechanics and Metallurgy, Third edition* (with R. M. Caddell); *Mechanical Behavior of Materials* (Cambridge); *Intermediate Materials Science* (Cambridge); *The Mechanics of Crystals and Textured Polycrystals*; *Physical Metallurgy*; and *Reporting Results* (with D. Van Aken; Cambridge).

Materials for Engineers

William F. Hosford

University of Michigan

CAMBRIDGE
UNIVERSITY PRESS

32 Avenue of the Americas, New York NY 10013-2473, USA

Cambridge University Press is part of the University of Cambridge.

It furthers the University's mission by disseminating knowledge in the pursuit of education, learning and research at the highest international levels of excellence.

www.cambridge.org
Information on this title: www.cambridge.org/9781107420519

First published 2008
First paperback edition 2014

A catalogue record for this publication is available from the British Library

Library of Congress Cataloguing in Publication data

Hosford, William F.
Materials for engineers / William F. Hosford.
    p. cm.
Includes bibliographical references and index.
ISBN: 978-0-521-89997-0 (hardback)
1. Materials – Textbooks.   I. Title.
TA403.H627   2008
620.1′1 – dc22   2008007734

ISBN 978-0-521-89997-0 Hardback
ISBN 978-1-107-42051-9 Paperback

Contents

Preface

The importance of materials to civilization is attested to by the names we give to various eras (Stone Age, Bronze Age, and Iron Age). We do not consider present times in terms of one specific material because so many are vital, including steel. The computer age would not be possible without silicon in computer chips.

Most introductory texts on materials science and engineering start with topics that are not of great interest to most engineers: atomic bonding, crystal structures, Miller indices. This introductory materials text differs from others because it is written primarily for engineers. It is shorter than most other materials texts so that it can easily be covered in one term. Emphasis is on mechanical and electrical properties of interest to most engineers. Thermal, optical, and magnetic behaviors are also covered. In addition, processing is treated in some detail.

Topics like X-ray diffraction, Miller indices, dislocations and coordination in compounds, surfaces, average molecular weights, Avrami kinetics, and Weibull analysis, which are of great interest to materials scientists but of little interest to most engineers, are covered only in the appendices. There is also an appendix on wood. There is no treatment of crystal systems, the Hall effect, or ferroelectricity.

After an introductory chapter, the text starts with phases and phase diagrams. This is followed by a chapter on diffusion, which treats diffusion in multiphase as well as single-phase systems. The next several chapters on mechanical behavior and failure should be of particular interest to mechanical engineers. There are chapters on iron and steel as well as nonferrous alloys. The chapters on electrical and optical properties should be of particular interest to electrical engineers. These are followed by chapters on specific types of materials: polymers, glasses, crystalline ceramics, clay products, forms of carbon, composite materials, concrete, and foams.

There is an emphasis on manufacturing, including recycling, casting and welding, powder processing, solid forming, and more modern techniques, including photolithography, vapor deposition, and the use of lasers. The text has been influenced by *Elements of Material Science* by L. H. Van Vlack and *Engineering Materials and Their Applications* by R. A. Flinn and P. J. Trojan. I want to thank David Martin for his help with polymers.

1 Introduction

Properties

One material is chosen over another for a particular application because its properties are better suited for the intended use. Among the important properties are strength, corrosion resistance, electrical conductivity, weight, material cost, processing costs, and appearance. Usually several properties are important.

In many applications, stiffness is important. Materials deform when a stress is applied to them. If the stress is low, the deformation will be elastic. In this case the deformation will disappear and the material will regain its original shape when the stress is removed. Good examples are elastic bands and paper clips. With greater stress, a material may deform plastically. In this case the deformation does not disappear when the stress is removed, so the shape change is permanent. This happens when the stress exceeds the yield strength of the material. With a still higher stress, the material may reach its tensile strength and fail. Ductility and toughness are also important. The ductility of a material is the amount of deformation a material may undergo before breaking. Toughness is a measure of how much energy a material absorbs per area before fracture.

Electrical properties are of paramount importance in some applications. Electrical conductors should have high conductivities and insulators very low conductivities. Integrated circuits for computers require semiconductors. The dielectric constant may be important in applications involving high frequencies. Piezoelectric behavior is required for sonar transmitters and receivers.

The corrosion resistance of metals in aqueous solutions is important in many applications. At high temperatures, resistance to direct oxidation is necessary. With polymeric materials, resistance to solvents and irradiation controls many applications.

All materials respond to magnetic fields to some extent. The magnetic properties of the few materials that have strong responses are discussed in Chapter 22.

Where weight is important, density must be considered. Surface appearance is often a crucial consideration. In almost all applications, material cost is of vital importance. Although prices of materials are usually quoted in price per weight ($/lb), weight per volume is usually more important. In addition, the cost of processing different materials is paramount.

Bonding

The bonds that hold materials together are classified as metallic, ionic, covalent, or van der Waals. In metals, individual atoms lose their valence electrons to form an "electron gas." This negatively charged electron gas holds the positively charged atoms together.

There is a transfer of valence electrons in ionic solids. Anions (metallic ions) lose valence electrons and become positively charged. Cations (non-metal ions) accept the extra electrons and become negatively charged. The electrostatic attraction bonds anions to cations.

Covalent bonding is the result of sharing electrons so the outer shells can be complete. For example, if a carbon atom with four valence electrons shares one electron with four neighboring carbon atoms, each carbon atom will have eight outer electrons. Bonding is usually partially ionic and partially covalent in character.

Van der Waals bonding is much weaker than metallic, ionic, and covalent bonding. It arises from the weak dipoles of molecules. Asymmetric molecules are likely to have dipoles, and bonding results from the attraction between the dipoles. Even symmetric molecules and atoms have statistical dipoles, which result in even weaker bonding. Hydrogen atoms in covalent molecules create strong dipoles. The term *hydrogen bonding* refers to van der Waals bonds resulting from these dipoles. Figure 1.1 schematically illustrates this.

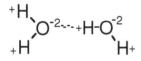

Figure 1.1. Hydrogen bond between adjacent water molecules resulting from the dipole of the hydrogen atom.

Table 1.1 gives the approximate ranges of the strengths of the different types of bonds. Bond strengths of organic compounds are given in Appendix 4.

Table 1.1. *Bond strengths*

Bond Type	Energy (kJ/mol)
Van der Waals	0.0–10
Hydrogen	10–40
Ionic	50–1000
Covalent	200–1000
Metallic	50–1000

Data from R. A. Flinn and P. T. Trojan, *Engineering Materials and Their Applications, 4th ed.* Houghton Mifflin, 1990.

Types of Materials

Most solid materials can be classified into three groups: metals, ceramics, and polymers.

In metals, the valence electrons are free to move about and are not bound to individual atoms. This explains their good electrical and thermal conductivities. It also explains the high reflectivity of metals. Photons are absorbed by promoting valence electrons to higher energy states and then re-emitting photons as the electrons drop to lower energy states. Metals are usually stiff and have high strengths but have enough ductility to be formed into useful shapes. Metals and alloys are crystalline. Aluminum and copper are typical metals. Brass (copper alloyed with zinc) and stainless steel (iron alloyed with chromium) are alloys.

The periodic table is useful in understanding some of the properties of metals. Densities correlate well with position in the periodic table, as shown in Figure 1.2. Figure 1.3 shows the melting points of various elements superimposed on the periodic table.

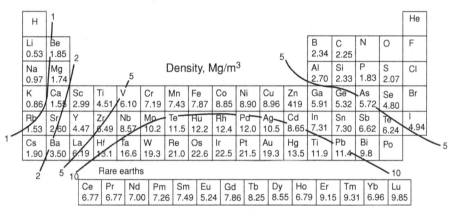

Figure 1.2. The heaviest elements are in the middle of the bottom of the periodic table. From W. F. Hosford, *Physical Metallurgy*, CRC, 2005.

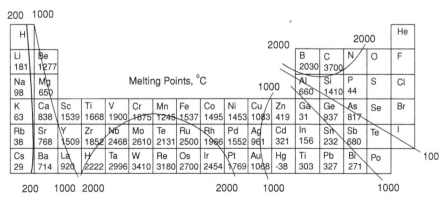

Figure 1.3. The metals with the lowest melting points are on the left of the periodic table, and those with the highest melting points, except boron and carbon, are in the middle at the bottom. From W. F. Hosford, *ibid.*

Ceramics are compounds of metals and non-metals. Oxides are particularly important. Because bonding is either ionic or covalent, or more likely partially ionic and partially covalent, electrons are not free to move, so most ceramics are electrical insulators. Ceramics with strong bonds are hard'and have very high melting points. Extremes are alumina, which is mostly covalent bonded, and sodium chloride, in which the bonding is mostly ionic. Ceramics may be either crystalline or amorphous. Most are stiff, hard, and brittle. Clay products and glasses are considered ceramics.

Polymers (commonly called plastics) consist of very large molecules that contain many repeating parts (*mers*). Weak van der Waals bonding between covalently bonded long chain molecules of thermoplastics explains their low stiffness and low melting points. Examples include polyethylene and polystyrene. In some polymers, all of the bonds are covalent and the molecules form three-dimensional networks. Examples include polyurethanes and bakelite. These polymers cannot be melted without decomposition. Silicones are polymers with silicon-oxygen backbones instead of carbon-carbon backbones.

Some materials, such as the various forms of carbon (graphite, diamond, and fullerenes), do not fit neatly into this classification. Graphite consists of sheets in which the bonding is metallic with only weak van der Waals bonding holding the sheets together. Diamond is similar to ceramics in that it is covalently bonded solid, but it consists of only one element. Bonding in semiconductors is similar to that in metals except that electrons are not free to move unless excited into a higher energy state.

Composites are mixtures of two or more different materials. Examples include glass fibers bonded by epoxy and graphite reinforcing polyester. Concrete is a composite of sand, gravel, and a cement paste. Foams and honeycombs may be regarded as composites of a solid with a gas.

Biological materials include wood, flesh, and bone. Flesh is an example of a soft biological material, and bone a hard biological material. Appendix 1 treats the structure and properties of wood.

Crystalline Materials

Materials can be classified as either crystalline or amorphous. In crystals, the atoms, ions, or molecules are arranged in three-dimensional patterns, described by unit cells, that repeat in space, thousands or millions of times in each dimension. Metals are almost always crystalline. There are several different crystal structures that are common in metals. In face-centered cubic (fcc) crystals, the atoms are on the corners of a cube and in the centers of the faces. Figure 1.4 shows the basic unit cell of an fcc crystal. Among the metals that have this structure are copper, silver, gold, aluminum, lead, nickel, and iron at elevated temperatures. Appendix 2 describes the system used to identify different planes and directions in a crystal, and Appendix 3 describes how X-ray diffraction can be used to determine crystal structures.

Example Problem 1–1

How many atoms are there in an fcc unit cell?

Solution: Counting each corner atom as 1/8 because it is shared by 8 cells and each face atom as 1/2 because it is shared by 2 cells, there are (8 corners)(1/8) + (6 faces)(1/2) = 4 atoms per units cell.

Figure 1.4. Face-centered cubic unit cell.

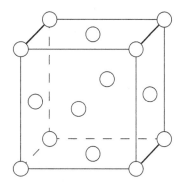

Another common metal structure is body-centered cubic (bcc). Atoms are on the corners and the center of a cubic cell (Figure 1.5). Iron at room temperature, chromium, vanadium, tungsten, and molybdenum are among the metals that have the bcc structure.

The third common crystal structure of metals is hexagonal close-packed (Figure 1.6). Magnesium, zinc beryllium, and zirconium and titanium (at room temperature) have this structure, which is composed of planes with atoms arranged in hexagonal patterns. The atoms in each hexagonal plane sit in the valleys of the plane below.

Many ceramics are crystalline. Bonding is usually partially ionic and partially covalent. They have crystal structures in which anions and cations are in contact. One of the simple structures is that of sodium chloride (Figure 1.7). Ceramic compounds that have this structure include MgO, FeO, and MnS. Figure 1.8 shows the unit cell of fluorite, CaF_2.

Figure 1.5. Body-centered cubic unit cell.

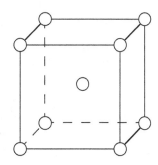

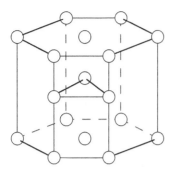

Figure 1.6. Close-packed hexagonal cell.

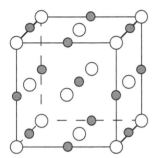

Figure 1.7. Unit cell of sodium chloride.

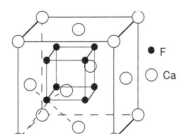

Figure 1.8. The unit cell of fluorite.

Amorphous Materials

Noncrystalline structures are said to be *amorphous*. Glasses are amorphous. Some polymers are partially crystalline and partially amorphous. Generally, stretching a polymer increases the degree of crystallinity by aligning its molecules. Polymers with three-dimensional networks are completely amorphous.

Amorphous materials have short-range order but not long-range order. Their structures are similar to liquids. Glasses have the structures of frozen liquids, but they are not liquids. Examples of amorphous materials include silicate glass (as in windows and bottles), most polymers, and even metallic alloys of complex compositions that have been cooled extremely rapidly from the melt.

Note of Interest

Johannes Diderik van der Waals was born on November 23, 1837, in Leyden, Netherlands. His first job was as a secondary school teacher. His lack of knowledge of the classic languages initially prevented him from enrolling at a university, until legislation removed this requirement for students in the sciences. His 1873 doctoral thesis "on the continuity of the gas and liquid state" demonstrated that the gas and liquid states merged. This work brought him to the attention of the scientific world. He continued with classic papers on binary surfaces and capillarity.

Problems

1. How many atoms are there in a body-centered cubic unit cell?
2. How many Na^+ ions are there in a unit cell of NaCl?
3. Knowing the bonding and structure of graphite, speculate as to why graphite is a good solid lubricant.
4. How many Ca^{2+} ions surround each F^- ion in the fluorite structure?
5. How many F^- ions surround each Ca^{2+} ion in the fluorite structure?

2 Phases

What Is a Phase?

A *phase* is a state of aggregation of matter that has a distinctive structure. Phases may be solid, liquid, or gaseous. A phase may be a pure material or a solution of several components. A solid phase is either amorphous or has a characteristic crystal structure and definite composition range. A physical system may contain more than one solid phase with different crystal structures, or different ranges of possible compositions – one or more mutually insoluble liquid phases (for example oil and water) and a gas phase. All gasses are mutually soluble, and therefore only one gaseous phase is possible.

Single-phase materials include brass (a solid solution of zinc in copper with zinc atoms occupying lattice sites), sodium chloride crystals, glass, and polyethylene. Most plain carbon steels are two-phase materials, consisting of an iron-rich solid solution and iron carbide. A glass of ice water consists of two phases: liquid water and ice. In a glass of tonic water there may be three phases: liquid, ice, and gas bubbles.

The composition of a single phase may vary from place to place, but composition change is always gradual and without abrupt changes. In multiple-phase systems, however, there are discontinuities of composition and structure at phase boundaries. The compositions on each side of the boundary are usually in equilibrium. For example, the oxygen concentration changes abruptly between copper with some oxygen dissolved in it and copper oxide.

Phase Diagrams

Phase diagrams describe the equilibrium between phases. Figure 2.1 shows the copper-silver phase diagram. Molten silver and copper form a single liquid in all proportions. When cooled, two different solid solutions form. There are two ways to consider this diagram. It may be thought of as a map, showing what phase or combination of phases exists at any temperature and composition. For example, in an alloy of 80% copper, a liquid and a copper-rich phase (β) exist at 850 °C. At 600 °C there

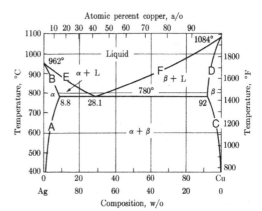

Figure 2.1. Copper-silver phase diagram.

are two solid phases – silver-rich α and copper-rich phase β. Above 1000 °C there is only one phase – a liquid.

A phase diagram may also be thought of as a plot of solubility limits. The lines A and B at the left of the diagram represent the solubility of copper in the silver lattice. The lines C and D at the right of the diagram represent the solubility of silver in the copper lattice. The lines E and F represent the solubility of silver and copper in the liquid.

At 600 °C, the solubility of copper in α-silver is about 3%, and the solubility of silver in β-copper is about 2.5%. All alloys in the two-phase α–β region will contain α saturated with silver and β saturated with copper. The compositions of two phases that are in equilibrium can be read by looking at the ends of a horizontal tie line connecting the two phases.

At 780 °C, three phases – liquid, α, and β – coexist, and their compositions are fixed. As liquid containing 28.1% Cu is cooled through 780 °C, the reaction, liquid $\rightarrow \alpha + \beta$, occurs. This is called a *eutectic* reaction, and 780 °C is called the eutectic temperature.

Lever Law

A simple mass balance can be used to find the relative amounts of each phase in a two-phase region. For example, in one gram of an alloy of 80% copper at 600 °C, there is 0.8 gram of copper, divided between the two phases. If we designate the fraction of the microstructure that is α by f_α, the amount of copper in the α-phase is $0.03 f_\alpha$. Likewise, there will be $0.975(1-f_\alpha)$ grams of copper in $(1-f_\alpha)$ grams of β-phase. A mass balance gives

$$0.8 = 0.03 f_\alpha + 0.975(1 - f_\alpha) \quad \text{or}$$
$$f_\alpha = (0.8 - 0.03)/(0.975 - 0.03) = 81.5\%.$$

This procedure can be generalized as the lever law

$$f_\alpha = (c_{\text{av}} - c_\alpha)/(c_\beta - c_\alpha). \qquad\qquad 2.1$$

Example Problem 2–1:

Consider an alloy containing 90% silver and 10% copper at 779 °C. What are the compositions of each phase, and what fraction of the alloy would be in the form of the silver-rich α-phase?

Solution: From Figure 2.1, the composition of the α-phase is 8.8% copper and the composition of the β-phase is 92% copper. Using the lever law, $f_\alpha = (92-10)/(92-8.8) = 96.7\%$.

Figures 2.2 through 2.4 are phase diagrams for several alloy systems. Copper and nickel are miscible in all proportions in both the liquid and solid states (Figure 2.2). Lead and tin form a simple eutectic (Figure 2.3). At the eutectic temperature (183 °C) $L \rightarrow \alpha + \beta$ on cooling. Figure 2.4 shows the copper-rich end of the copper-zinc phase diagram. In this system there is a *peritectic* reaction, $L + \alpha \rightarrow \beta$, on cooling through the peritectic temperature, 903 °C.

Probably the phase diagram with the greatest industrial importance is the iron-iron carbide phase diagram (Figure 2.5). The reaction $\gamma \rightarrow \alpha + Fe_3C$ at 727 °C is called a *eutectoid* reaction.

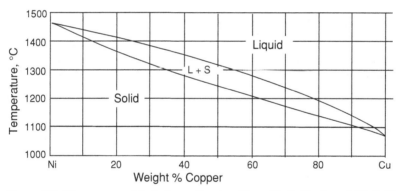

Figure 2.2. The copper-nickel phase diagram. From D. S. Clark and W. R Varney, *Physical Metallurgy for Engineers*, Van Nostrand Reinhold, 1962.

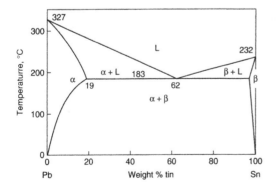

Figure 2.3. The lead-tin phase diagram. From W. F. Hosford, *Physical Metallurgy*, CRC, 2005.

Intermediate Phases

The β-phase in the Cu-Zn system is an example of an intermetallic solid solution. Although one might think of it as a compound CuZn, it differs from a conventional chemical compound in two important respects. It has a wide composition range, and the positions of the zinc and copper atoms are not fixed. They randomly occupy lattice sites in a body-centered cubic crystal structure. An alloy of β-brass is quite ductile.

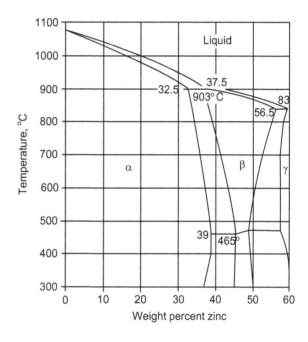

Figure 2.4. The copper-rich end of the copper-zinc phase diagram. From W. F. Hosford, *Physical Metallurgy*, CRC, 2005.

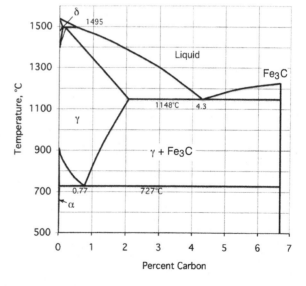

Figure 2.5. The equilibrium between iron and iron carbide.

Invariant Reactions

In binary systems equilibrium reactions involving three phases can occur only at fixed temperatures. The compositions of the phases are also fixed. Figure 2.6 illustrates four of the most common invariant reactions.

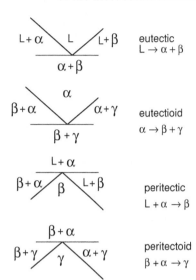

Figure 2.6. Four invariant reactions. The pertinent portion of the phase diagrams are shown at the left, and the reaction name and symbolic reaction on cooling at the right.

Effects of Pressure

Pressure, as well as temperature, affects equilibrium between phases. Figure 2.7 is a portion of the pressure-temperature phase diagram for water. Note that an increase of pressure tends to melt ordinary ice (phase Ih) because liquid water is denser than ice.

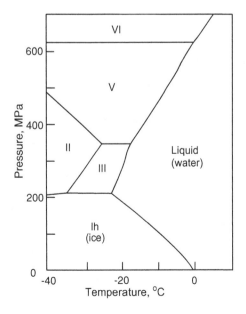

Figure 2.7. The pressure-temperature phase diagram for H_2O. The gas phase is not shown because it is stable only below a pressure of 0.6×10^{-3} MPa at $0\,°C$.

Solid Solutions

There are two types of solid solutions in metals. In *substitutional* solid solutions, solute atoms substitute for lattice atoms. In a system that has extensive solubility, the atoms have nearly the same size and valence.

In *interstitial* solid solutions, small atoms fit in the interstices between lattice atoms. Only small atoms can dissolve interstitially. Hydrogen is soluble in virtually all metals. Other elements that can dissolve interstitially are carbon, nitrogen, and boron. These have appreciable solubility only in transition metals. Both are illustrated in Figure 2.8.

Figure 2.8. Substitutional solid solution (left) and interstitial solid solution (right).

Lattice sites on which the atoms are missing are called *vacancies*.

Charge neutrality must be preserved in ionic crystals. A cation vacancy paired with an anion vacancy is called a *Schottky* defect. An example is formation of Li^+ and F^- vacancies in LiF. This is illustrated in Figure 2.9A. An anion vacancy may be paired with a nearby anion interstitial. This is called a *Frenkel* pair. An example is the formation of Zn^{2+} vacancies and Zn^{2+} interstitials in ZnO. This is illustrated in Figure 2.9B. In principle, paired cation vacancies and cation interstitials are possible,

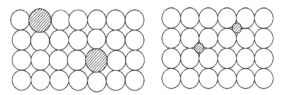

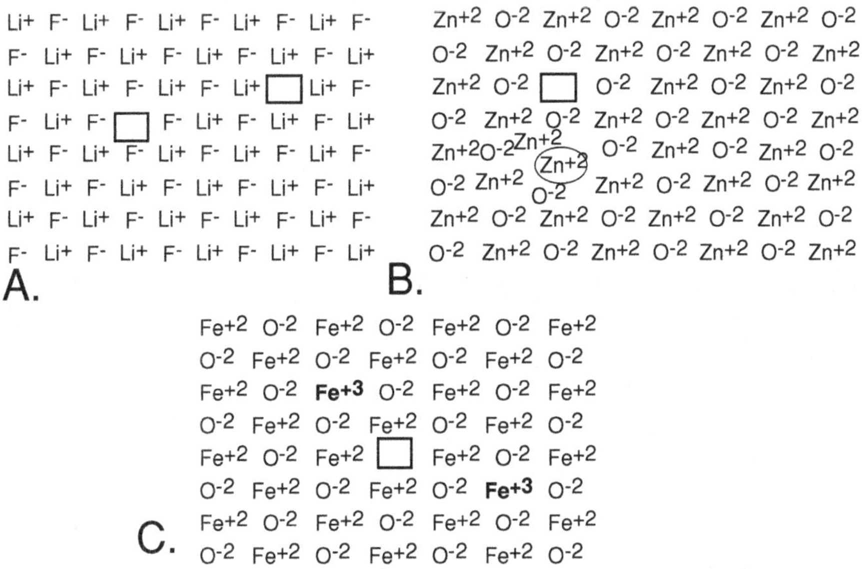

Figure 2.9. Shottky defects consist of paired cation and anion vacancies (A). Frenkel defects consist of ion vacancies and ion interstitial pairs (B). Anion vacancies may be neutralized by substitution of cations of higher valence (C).

but anion interstitials are unlikely because of the larger size of the anions. Solutions of cations that have a higher valence than that of the solvent cation are accompanied by anion vacancies. An example is the presence of Fe^{3+} ions in FeO. For every two Fe^{3+} ions, there is an O^{2-} vacancy shown in Figure 2.9C.

Example Problem 2–2:

A sample of wustite contains 77.6% iron and 22.4% oxygen by weight. What is the ratio of iron ions to oxygen ions? What fraction of the Fe^{2+} sites are vacant? What is the ratio of Fe^{3+} ions to O^{2-} ions?

Solution: $(77.6\,gFe/55.85\,g/mole\,Fe)/(22.4\,gO/16\,g/mole\,O) = 0.9925$. The fraction vacant Fe^{2+} sites is $1 - 0.9925 = 0.0075$. There must be one Fe^{3+} ion for every two vacant Fe^{2+} sites, so the ratio of Fe^{3+} ions to O^{2-} ions is 0.0038.

Grain Structure

Most crystalline materials are *polycrystalline*. That is, they are composed of many small crystals or grains with different orientations. There is disorder at the grain boundaries. Atoms take positions of minimum energy, but have fewer near neighbors at correct distances than atoms away from the grain boundary. Figure 2.10 is a schematic of the local structure of a grain boundary. In some places the atoms are too close, and in other places too far apart. Because of this the grain boundary has a characteristic energy. Solute atoms tend to segregate to grain boundaries. Atoms that are larger than the lattice atoms tend to substitute for atoms where they are too far apart, and small atoms tend to substitute where they are crowded. In that way they lower the grain boundary energy. The topic of surface energy is treated in Appendix 4.

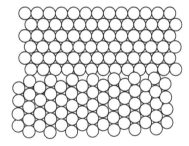

Figure 2.10. Atomic structure at a grain boundary. There are regions where the atoms are too far apart and others where they are too close.

Grain Size

There are two simple ways of characterizing the average size of grains. One is the linear intercept grain size. It is the average distance between grain boundaries on a random straight line drawn though the microstructure, corrected of course for magnification. The simple way of determining this is to divide the length of line by the number of intersections of the line with grain boundaries.

The other system is the ASTM grain size number, N, defined by

$$n = 2^{N-1},$$ 2.2

or

$$N = \ln n/\ln 2 + 1$$ 2.3

where n is the number of grains per square inch at a magnification of $100\times$.

Example Problem 2–3:

Figure 2.11 is a micrograph of an alloy at a magnification of $200\times$. Find the ASTM grain size.

Solution: There are 26 grains entirely within the field of view, 19 on the edge and 4 at the corners. Counting grains on the edge of the field as 1/2 and those on the corners as 1/4 in the field, there are $26 + 19/2 + 4/4 = 36.5$ grains. The area is $64\,\text{cm}^2/(2.54^2\,\text{cm}^2/\text{in}^2) = 9.92\,\text{in}^2$. Substituting $n = 36.5/9.92 = 3.7$ into equation 2.3, $N = \ln 3.7/\ln 2 + 1 = 2.9$.

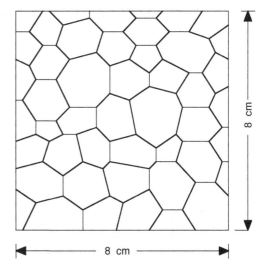

Figure 2.11. Micrograph of an alloy at a magnification of $200\times$.

8 cm

8 cm

Notes of Interest

Josiah Willard Gibbs (1839–1903) was an American theoretical chemist and physicist. He spent his entire professional career at Yale, which awarded him the first American PhD in engineering in 1863. His *On the Equilibrium of Heterogeneous Substances*, published in 1876, formed the basis for phase diagrams. As one of the greatest scientists of the nineteenth century, his work laid much of the theoretical foundation for chemical thermodynamics. He invented vector analysis. Although his work was little known during his life in America, it was widely recognized abroad. His work inspired Maxwell to personally make a plaster cast, which Yale still owns, illustrating Gibbs' construction.

Below 13.2 °C the equilibrium structure of tin is diamond cubic. However, the transformation of tin from the silvery and ductile *white* tin above 13.2 °C to the brittle *gray* tin is very sluggish. Because of a large volume expansion it eventually decomposes into powder, which is called tin pest. Tin cans over 80 years old have been discovered in Antarctica with the tin in good condition. Tin pest can be avoided by alloying with small amounts of antimony or bismuth.

Problems

1. Consider an alloy containing 50% nickel and 50% copper. Assume equilibrium.
   a. On cooling, what is the composition of the first solid to form?
   b. At 1300 °C, what are the compositions of the two phases?
   c. At 1300 °C, what fraction of the alloy is solid?
2. Consider Figure 2.7.
   a. Discuss the relative density of phases III and Ih (ice).
   b. Discuss the relative density of phase III and Liquid (water).
3. Consider Figure 2.12. At 120 °C, what fraction of the mixture of polybutadiene and polystyrene is in the form of Liquid 1?
4. If a steel containing 0.35% carbon is slowly cooled, what weight percent of the alloy will be in the form of Fe_3C? (See Figure 2.5.)

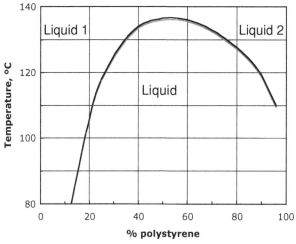

Figure 2.12. The polybutadiene-polystyrene phase diagram.

5. If the average diameter of grains in a metal was 1 mm, what would the ASTM grain size number be?

6. Name the invariant reactions that occur in the iron-iron carbide system at 727, 1148, and 1495 °C.

7. If the ratio of iron ions to oxygen ions in FeO is 0.99, what fraction of the Fe sites are filled with Fe^{3+} ions?

3 Diffusion

Fick's First Law

A concentration gradient in a phase creates a tendency for atoms to migrate in such a way as to eliminate the gradient, just as heat flows to even out temperature gradients. This tendency can be expressed by Fick's first law, which states

$$J = -D\,dc/dx, \qquad\qquad 3.1$$

where J is the net flux of solute, D is the diffusivity (or diffusion coefficient), c is the concentration of solute, and x is distance. See Figure 3.1.

The flux, J, is the net amount of solute crossing an imaginary plane per area of the plane and per time. The flux may be expressed as solute atoms/(m^{2}s), in which case the concentration, c, must be expressed as atoms/m^3. Alternatively, J may be expressed in terms of mass of solute in the units of (kg solute)/(m^{2}s) with c, expressed in (kg solute)/m^3. The diffusivity has dimensions of m^2/s and depends on the solvent, the solute, the concentration, and the temperature.[*]

Fick's first law for mass transport by diffusion is analogous to the laws of thermal and electrical conduction. For heat conduction, $q = k\,dT/dx$, where dT/dx is the thermal gradient (°C/m), k is the thermal conductivity, J/(ms°C), and q is the flux, J/(m^{2}s). Ohm's law, $I = E/R$, can be expressed in terms of a current density, $i = \sigma\varepsilon$, where i is the current density in coulombs/(m^{2}s) is a flux, ε is the voltage gradient (volts/m), and the proportionality constant is the conductivity, $\sigma = 1/\rho$, in (ohm·m)$^{-1}$.

Direct use of Fick's first law is limited to steady-state or nearly steady-state conditions in which the variation of dc/dx over the concentration range of concern can be neglected.

Fick's Second Law

Fick's second law expresses how the concentration at a point changes with time. Consider an element of area A and thickness dx in a concentration gradient

---

[*] W. C. Leslie, J. A. Gula, and A. A. Hendrickson, *American Mineralogist*, v. 20 p. 1067.

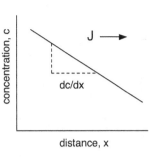

Figure 3.1. A concentration gradient. Note that with a negative gradient ($dc/dx < 0$) as shown here the flux, J, is positive. From W. F. Hosford, *Physical Metallurgy*, CRC, 2005.

(Figure 3.2). According to Fick's first law, the flux into the element is $J_{in} = -D\,dc/dx$ and the flux out of it is $J_{out} = -D\,dc/dx - \partial\,(-D\,dc/dx)\,\partial x$.

The rate of change of the composition within the element is $dc/dt = J_{in} - J_{out}$ so

$$dc/dt = \partial(D\,dc/dx)/\partial x. \qquad 3.2$$

This is a general statement of Fick's second law, which recognizes that the diffusivity may be a function of concentration and therefore of distance, x. In applications where the variation of D with distance and time can be neglected, equation 3.2 can be simplified into a more useful form,

$$dc/dt = D\,\partial^2 c/\partial x^2. \qquad 3.3$$

In principle concentration should be expressed in atoms or mass per volume, but if density changes are neglected concentration may be expressed in atomic % or weight %. Several specific solutions to Fick's second law are given below.

Solutions of Fick's Second Law

If the composition at the surface of a material is suddenly changed from its initial composition, c_0, to a new composition, c_s, and held at that level (Figure 3.3), the solution to equation 3.3 is

$$c = c_s - (c_s - c_0)\mathrm{erf}[x/(2\sqrt{Dt})], \qquad 3.4$$

where erf is the *error function*. Table 3.1 and Figure 3.4 show how $\mathrm{erf}(y)$ depends on y. One application of this solution involves carburizing and decarburizing of steels.

Three straight lines make a rough approximation to the error function as shown in Figure 3.5. For $y \geq 1$, $\mathrm{erf}(y) \approx 1$; for $1 \geq y \geq -1$, $\mathrm{erf}(y) \approx y$; for $-1 \geq y$, $\mathrm{erf}(y) \approx -1$.

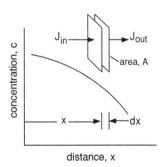

Figure 3.2. The rate of change of composition in an element of volume $A\,dx$ equals the differences between the fluxes into and out of the element. $\mathrm{erf}(y) -1$. For $y \geq 1$, $\mathrm{erf}(y) = 1$. For $-1 \leq y \leq 1$ $\mathrm{erf}(y) = y$.

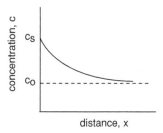

Figure 3.3. Solution to Fick's second law for a constant surface concentration, c_s. From W. F. Hosford, *Materials Science: An Intermediate Text*, Cambridge, 2006.

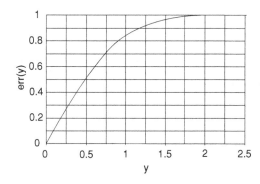

Figure 3.4. The dependence of erf(y) on y. W. F. Hosford, *Physical Metallurgy,* CRC, 2005.

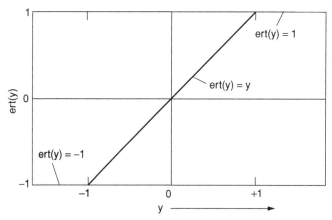

Figure 3.5. Approximation of the variation of erf(y) with y. For $y \leq -1$, erf(y) -1. For $y \geq 1$, erf(y) $= 1$. For $-1 \leq y \leq 1$, erf(y) $= y$. W. F. Hosford, *ibid.*

Junction of Two Solid Solutions

Another simple solution is for two blocks of differing initial concentrations, c_1 and c_2, that are welded together. In this case

$$c = (c_1 + c_2)/2 - [(c_1 - c_2)/2]\mathrm{erf}[x/(2\sqrt{Dt})]. \qquad 3.5$$

Figure 3.6 illustrates this solution. Note that equation 3.5 is similar to equation 3.4, except that $(c_1 + c_2/2)/2$ replaces c_s and $(c_1 - c_2/2)/2$ replaces $c_s - c_o$.

Table 3.1. *Values of the error function,* $\mathrm{erf} x = \frac{2}{\pi^{1/2}} \int \exp(-t^2) dt$

y	$\mathrm{erf}(y)$	y	$\mathrm{erf}(y)$	y	$\mathrm{erf}(y)$	y	$\mathrm{erf}(y)$
0.00	0.0000	0.05	0.0564	0.10	0.1125	0.15	0.1680
0.20	0.2227	0.25	0.2763	0.30	0.3286	0.35	0.3794
0.40	0.4284	0.45	0.4755	0.50	0.5205	0.55	0.5633
0.60	0.6039	0.65	0.6420	0.70	0.6778	0.75	0.7112
0.80	0.7421	0.85	0.7707	0.90	0.7970	0.95	0.8209
1.00	0.8427	1.10	0.8802	1.20	0.9103	1.30	0.9340
1.40	0.9523	1.50	0.9861	1.60	0.9763	1.70	0.9838
1.80	0.9891	1.90	0.9928	2.00	0.9953	2.20	0.9981
2.40	0.9903						

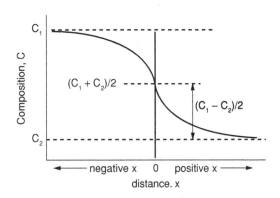

Figure 3.6. The solution to Fick's second law for two solutions with different concentrations. W. F. Hosford, *ibid.*

Temperature Dependence of Diffusivity

In substitutional solid solution, diffusion occurs by lattice vacancies interchanging positions with atoms as shown in Figure 3.7.

The equilibrium number of vacancies depends exponentially on temperature,

$$n_v = n_\mathrm{o}\exp(-E_\mathrm{f}/kT), \qquad\qquad 3.6$$

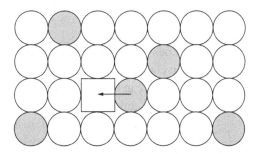

Figure 3.7. Diffusion in a substitutional solid solution occurs by interchange of lattice vacancies and atoms.

where n_v/n_o is the fraction of the lattice sites that are vacant and E_f is the energy to form a vacancy. The rate a given vacant site will be filled by a substitutional atom moving into it is also dependent on thermal activation,

$$\text{rate} = \exp(-E_m/kT), \qquad\qquad 3.7$$

where E_m is the energy barrier to fill a vacancy by movement of an adjacent substitutional atom. The net rate of diffusion is proportional to the product of the number of vacancies and the rate at which they contribute to diffusion. Therefore, $D = D_o\exp(-E_f/kT) \cdot \exp(-E_m/kT)$, which simplifies to

$$D = D_o\exp(-E/kT), \qquad\qquad 3.8$$

where

$$E = E_f + E_m. \qquad\qquad 3.9$$

Equation 3.8 can also be expressed in an equivalent form in terms of Q, the activation energy per mole of diffusion jumps,

$$D = D_o\exp(-Q/RT). \qquad\qquad 3.10$$

In interstitial solid solutions, diffusion occurs by movement of an interstitial atom from one interstitial site to another. Diffusion of interstitial atoms is much faster than diffusion of substitutional atoms.

Data for diffusion in several systems are given in Table 3.2.

Table 3.2. *Diffusion data for several metals*

Solute	Solvent	Q (kJ/mole$-°$C)	D_o (m^2/s)
Carbon	fcc iron	143	0.2×10^{-4}
Carbon	bcc iron	123	2.2×10^{-4}
Iron	fcc iron	269	0.22×10^{-4}
Iron	bcc iron	241	2.0×10^{-4}
Nickel	fcc iron	281	0.77×10^{-4}
Copper	aluminum	165	0.71×10^{-4}
Copper	copper	198	0.2×10^{-4}

Diffusion in Systems with More than One Phase

In analyzing diffusion couples involving two or more phases, there are two key points:

1. Local equilibrium is maintained at interfaces; therefore, there are discontinuities in composition profiles at interfaces. The phase diagram gives the compositions that are in equilibrium with each another.
2. No net diffusion can occur in a two-phase microstructure because both phases are in equilibrium, and there are no concentration gradients in the phases. These points will be illustrated by several examples.

Example Problem 3–1:

Consider diffusion between two pure metals in a system that has an intermediate phase as illustrated Figure 3.8. Interdiffusion between blocks of pure A and pure B at temperature T will result in the concentration profiles shown in Figure 3.9.

Note that a band of β develops between the α- and γ-phases. The compositions at the α–β and β–γ interfaces are those from the equilibrium diagram so the concentration profile is discontinuous at the interfaces. No two-phase microstructure will develop.

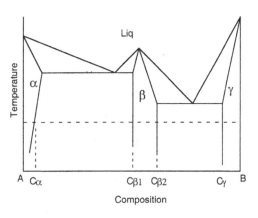

Figure 3.8. The A-B phase diagram. From W. F. Hosford, *Materials Science: An Intermediate Text*, Cambridge, 2006.

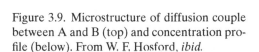

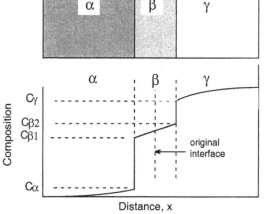

Figure 3.9. Microstructure of diffusion couple between A and B (top) and concentration profile (below). From W. F. Hosford, *ibid.*

Example Problem 3–2:

Consider the decarburization of a steel having a carbon content of c_0 when it is heated into the austenite (γ) region and held in air (Figure 3.10). At this temperature the reaction $2C + O_2 \rightarrow 2CO$ effectively reduces the carbon concentration at the

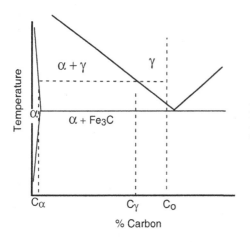

Figure 3.10. Iron-carbon diagram. From W. F. Hosford, *Materials Science: An Intermediate Text*, Cambridge, 2006.

surface to zero. A layer of α forms at the surface and into the steel to a depth x. The concentration profile near the surface is shown in Figure 3.11. The concentration gradient is $dc/dx = -c_\alpha/x$, where c_α is the carbon content of the α in equilibrium with the γ. Fick's first law gives the flux, $J = -dc/dx = Dc_\alpha$. As the interface advances a distance dx (Figure 3.12), the amount of carbon that is removed in a time interval dt is approximately $(c_\gamma - c_\alpha)dx$ so the flux is

$$J = (c_\gamma - c_\alpha)dx/dt. \qquad 3.11$$

Equating the two expressions,

$$(c_\gamma - c_\alpha)dx/dt = Dc_\alpha \qquad 3.12$$

so $xdx = D[c_\alpha/(c_\gamma - c_\alpha)]dt$. Integrating gives $x^2 = 2D[c_\alpha/(c_\gamma - c_\alpha)]$ or

$$x = [2Dc_\alpha/(c_\gamma - c_\alpha)]^{1/2}. \qquad 3.13$$

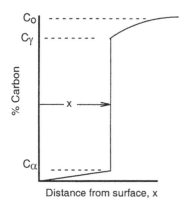

Figure 3.11. Carbon concentration profile. From W. F. Hosford, *ibid.*

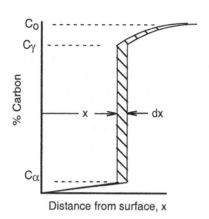

Figure 3.12. Change of concentration profile as the α–γ interface advances a distance, dx, in a time interval, dt. From W. F. Hosford, *Materials Science: An Intermediate Text*, Cambridge, 2006.

Example Problem 3–3:

It is known that with a certain carburizing atmosphere, it takes 8 hours at 900 °C to obtain a carbon concentration of 0.75 at a depth of 0.020 in. Find the time to reach the same carbon concentration at a depth of 0.30 in. at another temperature.

Solution: $x_2/\sqrt{D_2 t_2} = x_1/\sqrt{D_1 t_1}$. Letting $t_1 = 8$ hours, $x_1 = 0.020$ in., $x_2 = 0.035$ in., and $D_2 = D_1$. Then $t_2 = t_1(D_1/D_2)(x_2/x_1)^2 = (D_1/D_2)(0.035/0.02)^2 = 3.1(D_1/D_2)$.

Example Problem 3–4:

A steel containing 0.25% C was heated in air for 10 hours at 700 °C. Find the depth of the decarburized layer (i.e., the layer in which there is no Fe_3C). Given: The solubility of C in α-Fe at 700 °C is 0.016%. One may assume that the carbon concentration at the surface is negligible.

Solution: At 700 °C the steel consists of two phases, α and Fe_3C. The concentration profile must appear as shown in Figure 3.13. Near the surface there is a decarburized layer containing only α. The concentration in the α must vary from 0% C at the outside surface to $c_\alpha = 0.016\%$ C where it is in contact with Fe_3C. See Figure 3.10.

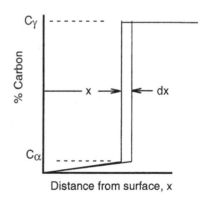

Figure 3.13. Decarburization of a steel heated in the $\alpha + Fe_3C$ phase region. From W. F. Hosford, *ibid.*

An approximate solution can be obtained by using Fick's first law to make a mass balance as the interface moves a distance dx. The amount of carbon transported to the surface in a period dt is $(\bar{c} - c_\alpha)dx$ and this must equal the flux times dt, $-Jdt = D$ $(dc/dx)dt$. Substituting $dc/dx = (c_\alpha - 0)/x$, $(\bar{c} - c_\alpha)dx = D(c_\alpha/x) \, dt$ and integrating, $x^2/2 = Dtc_\alpha/(\bar{c} - c_\alpha)$, $x = [2Dtc_\alpha/(\bar{c} - c_\alpha)]^{1/2}$. Now substituting $D = 2 \times 10^{-6} \exp[-84,400/(8.31 \times 973)] = 5.86 \times 10^{-11} \text{ m}^2/\text{s}$, $\bar{c} = 0.25$, $c_\alpha = 0.016$, and $t = 36,000\text{s}$, $x = 0.00057$ m or 0.6 mm.

Note that in all solutions of Fick's laws for fixed concentrations, $x/\sqrt{(Dt)}$ is constant or x is proportional to $\sqrt{(Dt)}$. Using this relation, many problems can be solved without use of the error function.

Notes of Interest

Adolf Eugen Fick was born September 3, 1829, in Kassel, Germany. He was a physician who, in 1887, invented the contact lens. In 1855 he proposed the law governing diffusion across a membrane.

In the Keewanaw Peninsula of Michigan there are nuggets, called *half breeds*, that contain both metallic copper and silver in intimate contact that were formed from aqueous solutions. A scanning electron microscope probe study[*] of the composition near the silver-copper boundary in half breeds indicates no detectable interdiffusion within the limits of resolution of the microscope beam. Using the relation $x = (tD)^{1/2}$, and assuming that the depth of diffusion was the maximum undetectable ($2 \, \mu\text{m}$) and that $D = 1.23 \times 10^{-4} (\text{m}^2/\text{s}) \exp[-194,000(\text{J/mol})/(RT)]$, it was concluded that if the deposit was 13,000 years old, its temperature could not have been higher than $300\,^\circ\text{C}$.

Problems

1. A block of copper containing 6% aluminum was welded to a block of copper containing 12% aluminum and heated to $800\,^\circ\text{C}$. Sketch how the concentration would vary after some diffusion had occurred. The copper-aluminum phase diagram is shown in Figure 3.14.

2. When iron was exposed to a carbon-bearing atmosphere at $850\,^\circ\text{C}$ for 2 hours, the carbon concentration 1.00 mm below the surface was found to be 0.65%. If the iron had been exposed to the same atmosphere at $900\,^\circ\text{C}$ for 1.5 hours, at what depth below the surface would the carbon concentration be 0.65%?

3. A block of silver containing 10% copper was cold welded to a block of copper containing 10% silver. It was then heated to $750\,^\circ\text{C}$. Sketch the concentration profile after a long period of time. See Figure 2.1 for the Ag-Cu phase diagram.

4. The concentration of carbon on the surface of iron containing 0.20% C is maintained at 1.00% while the iron is held at $900\,^\circ\text{C}$ for an hour. At what depth below the surface would the concentration be 0.45%?

* W. C. Leslie, J. A. Gula, and A. A. Hendrickson, *American Mineralogist*, 20: 1067.

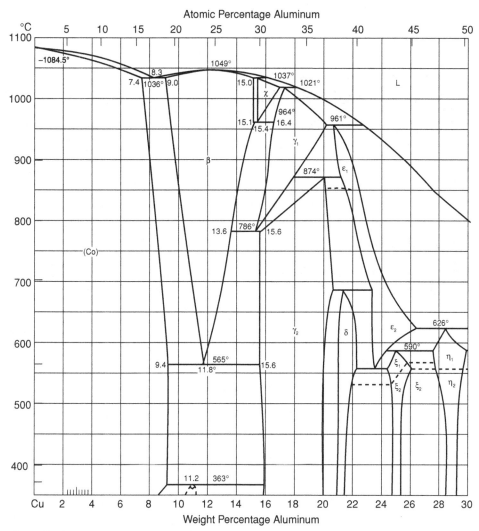

Figure 3.14. The copper-rich end of the copper-aluminum phase diagram. From American Society of Metals, *Metals Handbook*, 8th ed., v. 8, 1982.

4 Mechanical Behavior

Stress and Strain

Stress is defined as the intensity of force, F, per area. True stress, σ, is defined in terms of the current area, A,

$$\delta = F/A. \tag{4.1}$$

Nominal or engineering stress, S, in a tension or compression test is defined in terms of the original area, A_o,

$$S = F/A_o. \tag{4.2}$$

Strain describes the amount of deformation a material has undergone. An increment of true strain, $d\varepsilon$, is defined in terms of the length, L, as,

$$d\varepsilon = dL/L, \tag{4.3}$$

so the true strain is expressed as

$$\varepsilon = \ln(L/L_o), \tag{4.4}$$

where L_o is the original length.

Engineering or nominal strain, e, is defined simply as

$$e = \Delta L/L_o. \tag{4.5}$$

For small deformations, $e \approx \varepsilon$ and $S \approx \sigma$.

Elasticity

All materials deform when stressed. If the stress is small, the deformation is elastic. Elastic deformation is reversible so when the stress is removed the material will return to its original shape. A rubber band is a familiar example. Most materials, however, can undergo much less elastic deformation than rubber. The maximum elastic strain in crystalline materials is usually less than 0.2%. For most materials (not rubber) it is safe to assume that the strain is proportional to the stress. This assumption is the basis for the following treatment. Because elastic strains are small, it doesn't matter whether the elastic relations are expressed in terms of engineering strains, e, or true strains, ε.

An isotropic material is one that has the same properties in all directions. If uniaxial tension is applied in the x-direction, the tensile strain is $e_x = \sigma_x/E$ where E is *Young's modulus*. Uniaxial tension also causes lateral strains, $e_y = e_z = -\upsilon e_x$, where υ is *Poisson's ratio*. Consider the strain, e_x, produced by a general stress state, $\sigma_x, \sigma_y, \sigma_z$. The stress, σ_x, causes a contribution $e_x = \sigma_x/E$. The stresses, σ_y, σ_z, cause Poisson contractions, $e_x = -\upsilon\sigma_y/E$ and $e_x = -\upsilon\sigma_z/E$. Taking into account these Poisson contractions, the general statement of *Hooke's law* is

$$e_x = (1/E)[\sigma_x - \upsilon(\sigma_y + \sigma_z)]. \qquad\qquad 4.6a$$

Shear strains are affected only by the corresponding shear stress so

$$\gamma_{yz} = \tau_{yz}/G = 2e_{yz}, \qquad\qquad 4.6b$$

where G is the shear modulus. Similar expressions apply for all directions so

$$\begin{aligned} e_x &= (1/E)[\sigma_x - \upsilon(\sigma_y + \sigma_z)] & \gamma_{yz} &= \tau_{yz}/G \\ e_y &= (1/E)[\sigma_y - \upsilon(\sigma_z + \sigma_x)] & \gamma_{zx} &= \tau_{zx}/G \\ e_z &= (1/E)[\sigma_z - \upsilon(\sigma_x + \sigma_y)] & \gamma_{xy} &= \tau_{xy}/G. \end{aligned} \qquad 4.7$$

Table 4.1 lists values of Young's moduli for several materials.

Table 4.1. *Young's modulus at 20°C*

Material	Young's modulus (GPa)
Aluminum	70
Iron and steel	205
Copper	110
Magnesium	45
Titanium	116
Glass	70
Polystyrene	2.8
Nylon	2.8
LD polyethylene	0.1 to 0.35
Urea formaldehyde	10.3

Example Problem 4-1:

What stress is necessary to stretch an aluminum bar from 1.000 m to a length of 1.001 m?

Solution: The strain is $e = (1.001 - 1.000)/1.000 = 0.001$. Using the relation for uniaxial tension, $S = eE$, and taking $E = 70$ GPa from Table 4.1, $S = 0.001(70 \times 10^9)$ Pa $= 70$ MPa.

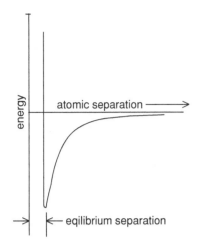

Figure 4.1. Schematic plot of the variation of binding energy with atomic separation. From W. F. Hosford, *Mechanical Behavior of Materials*, Cambridge, 2005.

Variation of Young's Modulus

The binding energy between two neighboring atoms or ions can be represented by a potential well. Figure 4.1 is a schematic plot showing how the binding energy varies with separation of atoms. At absolute zero, the equilibrium separation corresponds to the lowest energy. Young's modulus and the melting point are both related to the potential well; the modulus depends on the curvature at the bottom of the well, and the melting point depends on the depth of the well. The curvature and the depth tend to be related so the elastic moduli of different elements roughly correlate with their melting points. The modulus also correlates with the heat of fusion and the latent heat of melting, which are related to the depth of the potential well. For a given material, the elastic modulus decreases as the temperature is increased. In crystalline materials, the change of the modulus with temperature between absolute zero and the melting point is by no more than a factor of 5 (Figure 4.2). Because the modulus depends on the strength of near-neighbor bonds, small amounts of alloying

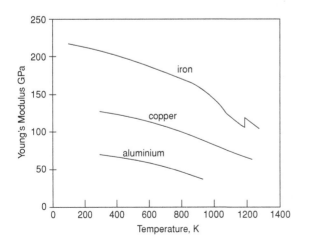

Figure 4.2. Decrease of Young's modulus with temperature. From W. F. Hosford, *ibid.*

elements and lattice defects caused by deformation have little effect on the modulus. It is, however, dependent on crystallographic direction.

The moduli of polymers, in contrast, are extremely temperature dependent. A change of temperature from 60 to 75 °C decreases the modulus of polyvinylchloride from 2 to 0.008 GPa as shown in Figure 4.3. As linear polymers are elongated, their molecules tend to align with the direction of elongation. This causes a large increase of the modulus. The amount of elastic deformation without permanent shape change is much greater in polymers than in crystals. Typically crystalline materials can be deformed elastically only a few tenths of a percent, whereas elastic deformation of a few percent is possible in polymers. In rubber, the elastic deformations are in the hundreds of percent. It should be noted, however, the elastic behavior is not linear (Hookean) at large strains.

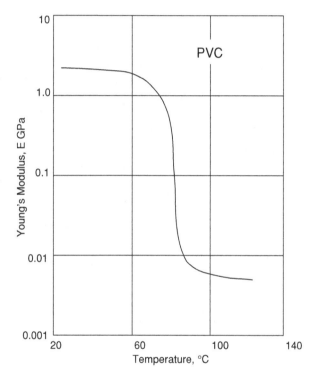

Figure 4.3. Temperature dependence of Young's modulus for polyethylene.

Plastic Deformation

If the deformation exceeds a critical amount, the material will not return to its original shape. This is illustrated schematically in Figure 4.4.

The material is said to have *yielded* and deformed *plastically*. The behavior is often characterized by the engineering stress-strain curve in tension. Figure 4.5 illustrates a specimen used in a tension test, and Figure 4.6 is an engineering stress-strain curve of a ductile material.

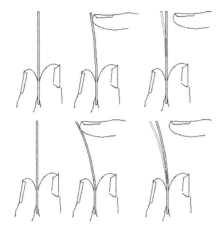

Figure 4.4. Use of the fingers to sense the elastic and plastic response of a wire. With a low force (top) the deformation is entirely elastic and the bending disappears when the force is removed. With greater force the elastic portion of the bending disappears when the force is removed, but some plastic deformation remains. From W. F. Hosford in *Tensile Testing*, ASM International, 1992.

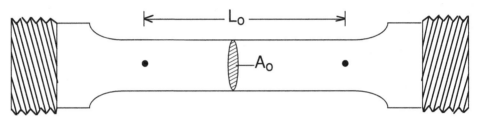

Figure 4.5. Typical tensile specimen.

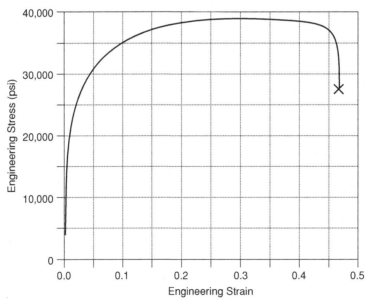

Figure 4.6. Typical engineering stress-strain curve for a ductile material. From W. F. Hosford, *Mechanical Behavior of Materials*, Cambridge, 2005.

It is tempting to define an *elastic limit* as the stress at which the very first plastic deformation occurs and a *proportional limit* as the stress at which stress is no longer proportional to strain; however, both of these quantities are sensitive to the accuracy of the stress and strain measurements. To avoid this problem, it is customary to define an *offset yield strength* that can be measured reproducibly. The offset yield strength is found by constructing a straight line parallel to the initial linear portion of the stress-strain curve, but offset from it by a strain of 0.002 or 0.2% as shown in Figure 4.7. The stress at the intersection of this line with the stress-strain curve is taken as the 0.2% offset yield strength. Loading a tensile specimen to this stress and then unloaded will result in a plastic strain of 0.002.

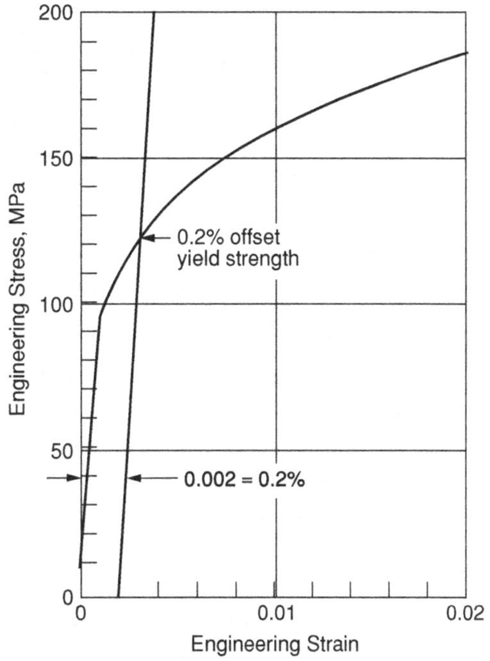

Figure 4.7. The first portion of a stress-strain curve illustrating how the 0.2% offset yield strength is determined. From W. F. Hosford, *ibid*.

An initial drop of stress occurs after the first yielding for a few materials. Examples include low-carbon steel (Figure 4.8A) and linear polymers (Figure 4.8B). For these materials an *upper yield strength* is defined as the stress at the first maximum and a *lower yield strength* is defined as the level of the plateau following initial yielding.

Example Problem 4–2:

The following data were taken during a tension test:

Strain	0.0005	0.001	0.0015	0.002	0.0025	0.003	0.004	0.005
Stress (MPa)	102	205	301	352	400	415	422	425

Determine the 0.2% offset yield strength.

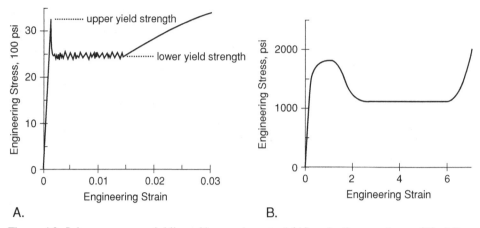

A.   B.

Figure 4.8. Inhomogeneous yielding of low-carbon steel (A) and a linear polymer (B). After the initial stress maximum, the deformation in both materials occurs within a narrow band that propagates the length of the gauge section before the stress rises again. From W. F. Hosford, *Mechanical Behavior of Materials*, Cambridge, 2005.

Solution: Plotting with an offset line parallel to the initial line and intersecting the strain axis at 0.002, the intersection is at 422 MPa.

The highest engineering stress is called the *tensile strength* or *ultimate strength*. If a material is ductile, as in Figure 4.6, the tensile strength is reached before the material breaks, and tensile strength corresponds to the point at which the deformation localizes into a neck as shown in Figure 4.9A. Figure 4.10 illustrates that necking starts at maximum engineering stress. For less ductile materials, fracture occurs before necking and in brittle materials before yielding, so the tensile strength is the fracture strength (Figure 4.9A and B).

The ductility of a material describes the amount of deformation before fracture in a tension test. There are two common measures of ductility. One measure is the *percent elongation* before fracture,

$$\% \, El = (L_f - L_o)/L_o \times 100\%. \qquad\qquad 4.8$$

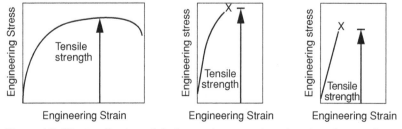

Figure 4.9. The tensile strength is the maximum engineering stress in a tension test, regardless of whether the specimen necks, fractures before necking, or fractures before yielding. From W. F. Hosford, *ibid*.

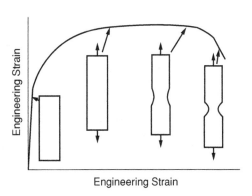

Figure 4.10. Necking starts when the maximum engineering stress is reached. From W. F. Hosford, *Mechanical Behavior of Materials*, Cambridge, 2005.

The other is the *percent reduction of area*,

$$\%RA = (A_o - A_f)/A_o \times 100\%. \qquad\qquad 4.9$$

The percent reduction of area is a better measure of ductility because the percent elongation involves both the uniform elongation before necking and the elongation during necking, and the latter depends on the specimen diameter. Tensile specimens with large length-to-diameter ratios have lower percent elongations than those with low length-to-diameter ratios.

It should be realized that the amount of deformation that a material can withstand depends on the form of loading. Under compression, as in rolling, the ductility is far higher than that measured in a tension test.

Strength Dependence on Grain Size

Grain boundaries are obstacles to deformation, so materials with finer grain sizes tend to have higher yield strengths. The Hall-Petch equation relates yield strength, σ_y, to the intercept grain size:

$$\sigma_y = \sigma_0 + K_Y d^{-1/2}, \qquad\qquad 4.10$$

where σ_0 and K_Y are constants and d is the average grain diameter. Figure 4.11 shows a plot of the Hall-Petch equation for low-carbon steel.

Example Problem 4–3:

The following data have been reported for the yield strength of low-carbon steel of two grain sizes.

Grain size (mm)	0.025	0.15
Yield strength (MPa)	205	125

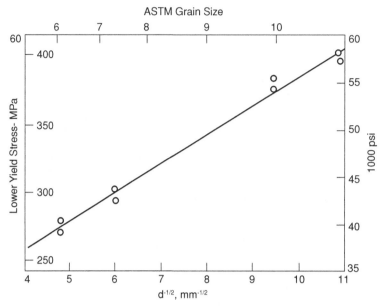

Figure 4.11. The yield strength of a low-carbon steel is inversely proportional to $\sqrt{d}$ as predicted by the Hall-Petch equation. Data from W. B. Morrison, *J. Iron and Steel Inst.*, v. 201, adapted from R. A. Grange, *ASM Trans.*, v. 59, 1966. W. F. Hosford, *Mechanical Behavior of Materials*, Cambridge, 2005.

Estimate the yield strength of the same steel heat-treated to give a grain size of 0.015 mm.

Solution: Since $\sigma_y - \sigma_o + K_y D^{-1/2}$, $(\sigma_y)_1 - (\sigma_y)_2 = K_y(D_1^{-1/2} - D_2^{-1/2})$; $K_y = [(\sigma_y)_1 - (\sigma_y)_2]/(D_1^{-1/2} - D_2^{-1/2}) = 80/(8.1165 - 6.325) = 21.4$. Substituting in $\sigma_y = \sigma_o + K_y D^{-1/2}$, $\sigma_o = \sigma_y - K_y D^{-1/2} = 205 - 21.4(0.025^{-0.5}) = 69.8$.

For $D = 0.015$, $\sigma_y = 69.8 + 21.4(0.015^{-1/2}) = 244$ MPa.

Solid Solution Strengthening

Elements in solid solution usually increase yield and tensile strengths. Brass is an alloy of copper and zinc. Figure 4.12 shows that zinc causes a large increase of the tensile strength. Strength also increases with finer grain sizes.

Cold Work

Just as the plastic deformation in a tension test increases the stress for continued deformation, plastic deformation by cold rolling, drawing, or any other means has the same effect. This is called *work hardening* or *strain hardening*. It is the principle means by which single-phase alloys can be hardened. Figure 4.13 shows the effect

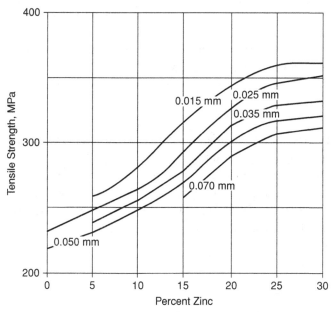

Figure 4.12. Copper is strengthened by zinc in solid solutions and by finer grain sizes. From *Understanding Copper Alloys*, J. H. Mendenhall, ed., Olin Brass, 1977.

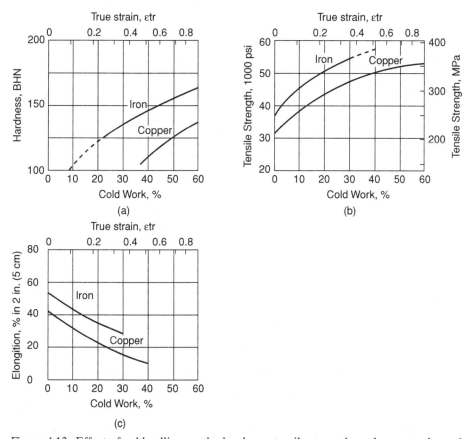

Figure 4.13. Effect of cold rolling on the hardness, tensile strength, and percent elongation in tension on iron and copper. From L. H. Van Vlack, *Elements of Materials Science and Engineering*, 6th ed., Addison-Wesley, 1989.

of cold rolling on the hardness, tensile strength, and percent elongation in a tension test of iron and copper after various amounts of cold work.

Hardness

The hardness test is a simple way of characterizing a material's strength. An indenter is pressed into the surface of a material under a fixed force and the extent of indentation measured. With the Brinell, Vickers, and Knoop tests, the dimensions of the indentation are measured after the load is removed, and the hardness is the load divided by the area of the indentation. With the several Rockwell tests, the depth of the indentation is automatically recorded as the hardness number. Figure 4.14 illustrates several hardness tests.

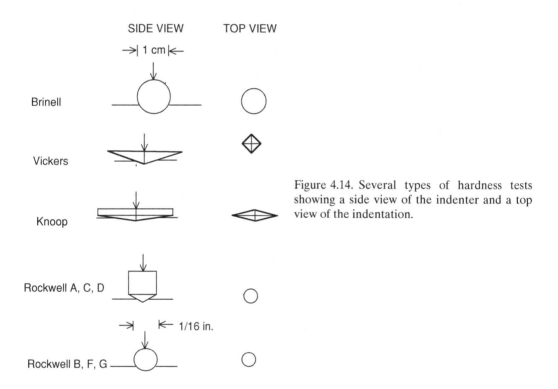

Figure 4.14. Several types of hardness tests showing a side view of the indenter and a top view of the indentation.

Slip

The primary mechanism by which plastic deformation occurs is called *slip*. Slip is the sliding of atomic planes over each other. Both the planes on which slip occurs and the directions of slip are crystallographic. In metals, the planes on which slip occurs are the closest-packed planes, and the directions of slip are the directions with shortest repeat distance. Figure 4.15 shows slip lines in aluminum.

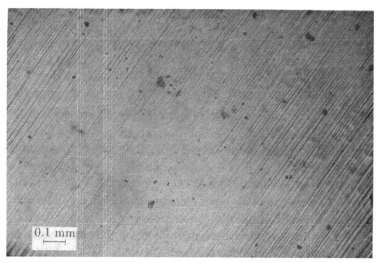

Figure 4.15. Slip lines on the surface of an aluminum grain. From F. A. McClintock and A. S. Argon, *Mechanical Behavior of Materials*, Addison-Wesley, 1966.

Slip occurs by the movement of dislocations, which are line defects in a crystal. See Appendix 5.

Notes of Interest

Thomas Young, who first described elastic moduli, was born in June 1773. His main work was in optics, physiology, and Egyptology. His research on optics contributed to the theory of the wave nature of light. His experiment of passing light through two parallel slits to form bands on a surface led him to reason that light was composed of waves. In 1793, he explained how the eye can focus at different distances by changing the curvature of its lens. In 1801, he first described astigmatism and postulated that the eye perceives color with three kinds of nerves in the retina. This was experimentally proven in 1959.

The Brinell hardness test was proposed in 1900 by Swedish engineer Johan August Brinell.

The yield point effect in linear polymers may be experienced by pulling the piece of plastic sheet that holds a six-pack of carbonated beverage cans together. When one pulls hard enough the plastic will yield and the force drop. A small thinned region develops. As the force is continued, this region will grow. The yield point effect in low-carbon steel may be experienced by bending a low-carbon steel wire. (Florist's wire works well.) First heat the wire in a flame to anneal it. Then bend a six-inch length by holding it only at the ends. Instead of bending uniformly, the deformation localizes to form several sharp kinks. Why? Bend an annealed copper wire for comparison.

Problems

1. A specimen was stretched from 1.051 in. to 1.052 in. Calculate the engineering strain and the true strain. What is the percent difference?

2. If a piece of steel ($E = 205$ GPa and $\upsilon = 0.28$) were elastically stretched in such a way that the strain in the x direction $e_x = 0.0012$ and the strain in the y-direction, $e_y = 0.00$, and the stress, $S_z = 0$, what stress, S_x, would be required in the x-direction?

3. Using the data in Figure 4.12, determine the constants in equation 4.10 for a brass containing 15% Zn.

4. In a tension test, a material fractured before necking occurred. The true stress and true strain at fracture were 600 MPa and 0.14, respectively. What is the tensile strength of the material?

5. Assuming constant volume and uniform elongation, find the relation between percent elongation and percent reduction of area.

6. The results of a tensile test on a steel test bar are given as follows. The initial gauge length was 25.0 mm and the initial diameter was 5.00 mm. The diameter at the fracture was 2.6 mm. The engineering strain and engineering stress in MPa are:

strain	stress	strain	stress	strain	stress
0.0	0.0	0.06	319.8	0.32	388.4
0.0002	42.	0.08	337.9	0.34	388.0.
0.0004	83.	0.10	351.1	0.38	386.5
0.0006	125.	0.15	371.7	0.40	384.5
0.0013	170.	0.20	382.2	0.42	382.5
0.005	215.	0.22	384.7	0.44	378.
0.02	249.7	0.24	386.4	0.46	362.
0.03	274.9	0.26	387.6	0.47	250.
0.04	293.5	0.28	388.5		
0.05	308.	0.30	388.5		

a. Plot the engineering stress-strain curve.
b. Determine
  i.   Young's modulus.
  ii. the 0.2% offset yield strength.
  iii. the tensile strength.
  iv. the percent elongation.
  v.   the percent reduction of area.

7. In a tension test with a 1.25 cm diameter bar with a 5 cm gauge length of an aluminum alloy, the percent reduction of area was found to be 50% and the percent elongation was 35%. In a tension test on a specimen of the same material with a 0.20 cm diameter bar and a 5 cm gauge length, the percent reduction of area was

50% and the percent elongation was 15%. Explain the differences in percent elongation.

8. In one tension test the neck formed outside of the gauge section. How did this affect the measured percent elongation?

9. Discuss whether increasing the hardness of a steel will increase its stiffness.

5 Mechanical Failure

Fracture

Typically as the yield strength of a material is increased, its ductility and toughness decrease. *Toughness* is the energy absorbed in fracturing. If a material has a high yield strength, it can be subjected to stresses high enough to cause fracture before there has been much plastic deformation to absorb energy. Factors that inhibit plastic flow lower ductility as schematically indicated in Figure 5.1. These factors include decreased temperatures, increased strain rates, and the presence of notches.

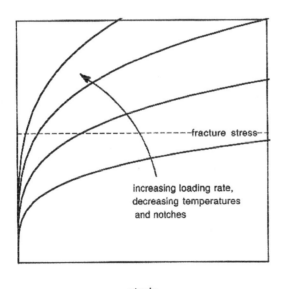

Figure 5.1. Lowered temperatures, increased loading rates, and the presence of notches all reduce ductility. These three factors raise the stress level required for plastic flow so the stress required for fracture is reached at lower strains. From W. F. Hosford, *Mechanical Behavior of Materials*, Cambridge, 2005.

Engineers should be interested in ductility and fracture for two reasons. Ductility is required to form metals into useful parts by forging, rolling, extrusion, or other plastic working processes. Also, some plastic deformation is necessary to absorb energy so as to prevent failure in service.

Fractures can be classified several ways. A fracture may be described as being *ductile* or *brittle*, depending on the amount of deformation that precedes it. Failures may also be described as *intergranular* or *transgranular*, depending on the fracture path. The terms *cleavage*, *shear*, *void coalescence*, etc., are used to identify failure mechanisms. These descriptions are not mutually exclusive. A brittle fracture may be intergranular, or it may occur by cleavage.

The failure in a tensile test of a ductile material occurs well after the maximum load is reached and a neck has formed. In this case, fracture usually starts by nucleation of voids in the center of the neck where the hydrostatic tension is the greatest. As deformation continues, these internal voids grow and eventually link up by necking of the ligaments between them (Figure 5.2). With continued elongation, this internal fracture grows outward until the outer rim can no longer support the load. At this point the edges fail by sudden shear. This overall failure is often called a *cup and cone* fracture. If the entire shear lip is on the same broken piece, it forms a cup. More often, however, part of the shear lip is on one half of the specimen and part on the other half. Figure 5.3 is a photograph of a cup and cone fracture.

In ductile fractures, voids form at inclusions because either the inclusion-matrix interface or the inclusion itself is weak. Figure 5.4 shows the fracture surface formed by coalescence of voids. Inclusions can be seen in some of the voids. The inclusion

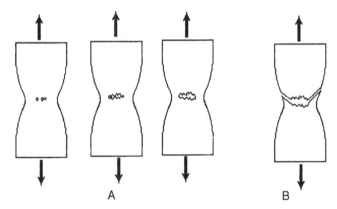

A   B

Figure 5.2. Development of a cup and cone fracture. (A) Internal porosity growing and linking up. (B) Formation of a shear lip.

Figure 5.3. A typical cup and cone fracture in a tension test of a ductile manganese bronze. From A. Guy, *Elements of Physical Metallurgy*, Addison-Wesley, 1959.

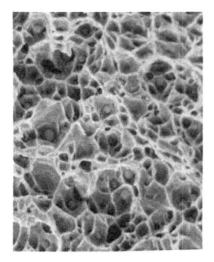

Figure 5.4. Dimpled ductile fracture surface in steel. Note the inclusions associated with about one half of the dimples. The rest of the inclusions are on the mating surface. Courtesy of J. W. Jones. From W. F. Hosford and R. M. Caddell, *Metal Forming: Mechanics and Metallurgy*, 3rd ed., Cambridge, 2007.

content of the material strongly influences its ductility. With increasing numbers of inclusions, the distance between the voids decreases so it is easier for them to link together and lower the ductility. Figure 5.5 shows the decrease of ductility with increasing amounts of inclusions.

Brittle fractures often occur by cleavage on specific atomic planes. Figure 5.6 is a schematic drawing of a cleavage fracture. In three dimensions the cleavage planes in one grain of a polycrystal will not link up with cleavage planes in a neighboring grain, so fracture of polycrystalline material cannot occur totally by cleavage. Another mechanism must link up the cleavage fractures in different grains. Figure 5.7 is as picture of an actual cleavage fracture.

Some polycrystals have brittle grain boundaries that form easy fracture paths. Figure 5.8 shows such an *intergranular fracture* surface. The brittleness of grain

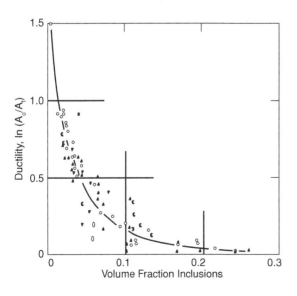

Figure 5.5. Effect of second-phase particles on the tensile ductility of copper. Data include alumina, silica, molybdenum, chromium, iron, and iron-molybdenum inclusions as well as holes. From B. I. Edelson and W. M. Baldwin, *Trans. Q. ASM.*, v. 55, 1962.

Figure 5.6. In polycrystalline material, cleavage planes in neighboring grains are tilted by different amounts relative to the plane of the paper; therefore, they cannot be perfectly aligned with each other. Another mechanism is necessary to link up the cleavage fractures in neighboring grains.

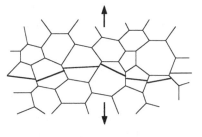

Figure 5.7. Cleavage fracture in an Fe-3.9%Ni alloy. The arrow indicates the direction of crack propagation. From American Society for Metals, *Metals Handbook*, 8th ed., v. 9, 1974.

SEM fractograph

Figure 5.8. Grain boundary fracture.

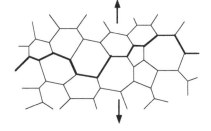

boundaries may be inherent to the material, or may be caused by segregation of impurities to the grain boundary or even by a film of a brittle second phase. Commercially pure tungsten and molybdenum fail by grain boundary fracture. These metals are ductile only when all the grain boundaries are aligned with the direction of elongation, as in tension testing of cold-drawn wire. Copper and copper alloys are severely embrittled by a very small amount of bismuth, which segregates to and wets the grain boundaries. Molten FeS in the grain boundaries of steel at hot-working temperatures causes failure along grain boundaries. Such loss of ductility at high temperatures is called *hot shortness*. Hot shortness of steel is prevented by adding Mn, which reacts with the sulfur to form MnS. Manganese sulfide is not molten at hot-working temperatures and does not wet the grain boundaries. Stress corrosion is responsible for some grain boundary fractures.

Creep

At high temperatures materials undergo creep which is time-dependent plastic deformation. Figure 5.9 shows typical creep behavior under constant load. There is an instantaneous strain on application of stress. This is followed by transient creep, during which the creep rate decreases over time. In stage II, the creep rate is almost constant and finally the creep rate increases to failure during stage III. The acceleration during stage III is a result of the ever-decreasing cross-sectional area increasing the true stress on the specimen. Porosity may develop during this stage.

The rate of creep increases with increasing stress and increasing temperature as shown in Figure 5.10. It is worth noting that the strain to failure usually decreases

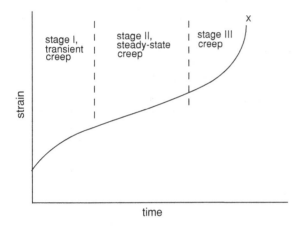

Figure 5.9. Typical creep curve. From W. F. Hosford, *Mechanical Behavior of Materials*, Cambridge, 2005.

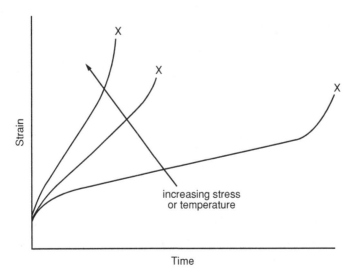

Figure 5.10. Decreasing temperature and stress lead to slower creep rates, but failure often occurs at a lower strain. W. F. Hosford, *Mechanical Behavior of Materials*, Cambridge, 2005.

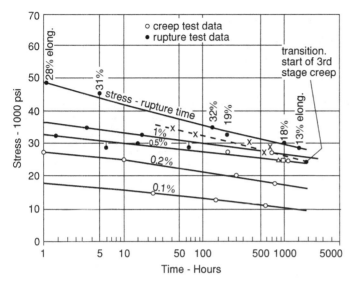

Figure 5.11. Stress dependence of stress-rupture life and time to reach several creep strains for a Ni-Cr-Co-Fe alloy tested at 650 °C. Note that as the stress level is reduced to increase the time to a given strain, rupture occurs at lower strains. From N. J. Grant in *High Temperature Properties of Metals*, ASM, 1950.

as the temperature and stress are lowered to achieve lower rates of creep. This is illustrated in Figure 5.11. Failure in service more often occurs because of rupture than excessive creep deformation.

Fatigue

Fracture can occur when the material is repeatedly stressed at levels well below that which would cause failure in a single loading. This is called *fatigue*. Even with loads below the yield strength, some plastic deformation occurs on each application of load. The fatigue behavior of a material is often represented by an *S-N* curve (*S* for stress and *N* for number of cycles to failure). Figure 5.12 shows the *S-N* curve for a steel. There is a stress level below which the steel can withstand an infinite number of cycles without failure. This is called the *fatigue limit* or *endurance limit*, and probably is the level of stress below which there is no plastic deformation. With steels this is usually about half of the tensile strength; however, most materials do not have a fatigue limit. The *S-N* curve for aluminum (Figure 5.13) is typical.

A number of factors govern the fatigue strengths of materials. These include average stress level, stress concentration, surface finish, residual stresses in the surface, temperature, and the presence of stress concentrators.

Cyclic stresses are usually applied about some mean stress as shown in Figure 5.14. The permissible cyclic stress decreases with increasing mean stress. Figure 5.15 shows several approximations to the effect. According to Goodman,

$$\sigma_a = \sigma_e[1 - (\sigma_m/TS)^2], \qquad\qquad 5.1$$

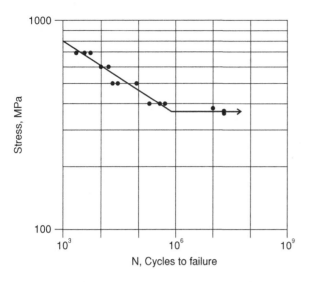

Figure 5.12. Typical *S-N* curve for a steel with a fatigue strength (or endurance limit) below which failure never occurs. From W. F. Hosford, *Mechanical Behavior of Materials*, Cambridge, 2005.

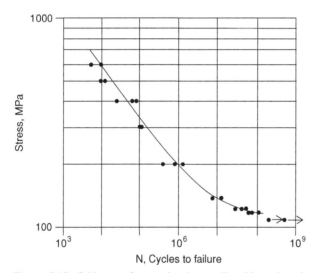

Figure 5.13. *S-N* curve for an aluminum alloy. Note that there is no true fatigue limit. This is typical of most metals. From W. F. Hosford, *ibid.*

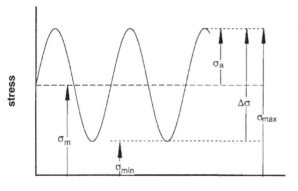

Figure 5.14. Schematic of cyclic stresses illustrating several terms. From W. F. Hosford, *ibid.*

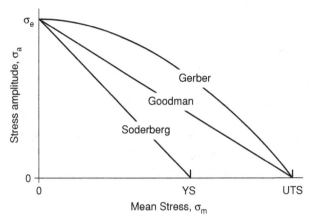

Figure 5.15. The effect of mean stress on the cyclic stress to give the same life as σ_e according to several approximations. From W. F. Hosford, *ibid.*

where σ_a is the stress amplitude corresponding to a certain life, σ_m is the mean stress, and σ_e is the stress amplitude that would give the same life if σ_m were zero. *TS* is the tensile strength. Soderberg proposed a more conservative relation:

$$\sigma_a = \sigma_e[1 - \sigma_m/YS], \qquad 5.2$$

where *YS* is the yield strength. Gerber proposed

$$\sigma_a = \sigma_e[1 - (\sigma_m/TS)^2]. \qquad 5.3$$

All three predict that the permissible stress amplitude decreases with a mean tensile stress.

The mean stress is changed by the presence of residual stresses. Because fatigue failures almost always initiate at the surface, a residual stress pattern with tension on the surface lowers the life.

Example Problem 5–1:

A steel has an endurance limit of 350 MPa, a yield strength of 450 MPa, and a tensile strength of 700 MPa. If it is under a steady-state stress of 50 MPa, what is the allowable cyclic amplitude according to the Goodman relation? The Soderberg relation?

Solution: Using the Goodman relation, $\sigma_a = \sigma_e(1-\sigma_m/UTS) = 350(1-50/700) = 325$ MPa.

Using the Soderberg relation, $\sigma_a = \sigma_e(1-\sigma_m/YS) = 350(1-50/450) = 311$ MPa.

Fatigue behavior is improved if the surfaces are left under residual compression and worsened if they are left under residual tension. This is consistent with the effects of mean stress. If the surface is in residual compression, cycling will be about a negative mean stress. Often parts that may be subjected to cyclic loading are shot peened to induce residual compression in the surface.

The *S-N* curve describes the fatigue behavior for a cyclic stress of constant amplitude; however, the stress cycles are often variable. The springs of a typical automobile are subjected to a few very large stress cycles when the car hits a pothole, and many lower cycles in ordinary driving. Palmgren and Miner suggested that a simple approximation can be used to analyze the fatigue life under these conditions. They proposed that failure occurs when

$$n_1/N_1 + n_2/N_2 + n_3/N_3 + \cdots = 1, \qquad\qquad 5.4$$

where n_1, n_2, and n_3 are the number of cycles at stress levels 1, 2, and 3, and N_1, N_2, and N_3 are the number of cycles that would cause failure for cycling at constant stress amplitudes 1, 2, and 3. This is called the Palmgren-Miner rule.

Example Problem 5–2:

According to Figure 5.13 the fatigue life of the aluminum is 6×10^6 cycles at 150 MPa, 10^6 cycles at 200 MPa, 1.5×10^5 cycles at 300 MPa, and 5×10^4 cycles at 400 MPa. If the aluminum is subjected to 10^6 cycles at 150 MPa, 4.8×10^5 cycles at 200 MPa, and 1.7×10^4 cycles at 300 MPa, how many cycles could it withstand at 400 MPa?

Solution: According to the Palmgren-Miner relation, the fraction of the life already expended is $10^6/6 \times 10^6 + 4.8 \times 10^5/10^6 + 1.5 \times 10^4/5 \times 10^4 = 0.947$. The remaining life at 400 MPa is $(1-0.947)5 \times 10^4 = 265 \times 10^3$ cycles.

Notches and abrupt changes in cross section lead to stress concentration. The actual stress at the root of a notch may be much larger than the nominal stress. The sharper the notch, the higher the stress concentration. In materials with low yield strengths, the problem is less severe because plastic deformation may make the notch less sharp. The roughness of the surface has a large effect on fatigue lives as shown in Figure 5.16. Corrosive environments also lower fatigue strengths.

Increased temperatures are detrimental because they lower the stress at which plastic deformation can occur, and therefore lower the fatigue strength.

In the very early stages of fatigue, slip causes extrusions and intrusions that act as stress concentrators. Figure 5.17 shows intrusions and extrusions caused by slip on different planes during the tension and compression portions of loading. Figure 5.18 is a schematic illustrating this.

Examination of a fatigue fracture surface often reveals *clamshell* or *beach markings* that show the progression of the fracture (Figure 5.19).

At a high magnification striations that show the cycle-to-cycle progression of a fracture may be seen (Figure 5.20).

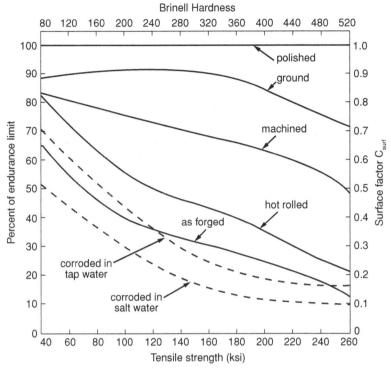

Figure 5.16. Effect of surface finish on fatigue behavior of steel. From C. Lipson and R. C. Juvinal, "Application of Stress Analysis to Design and Metallurgy," The University of Michigan Summer Conference, Ann Arbor, MI, 1961.

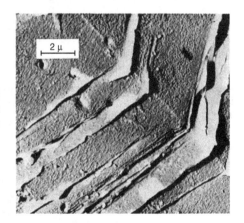

Figure 5.17. Surface observation of intrusions and extrusions caused by cyclic deformation. From A. Cottrell and D. Hull, *Proc. Roy. Soc. (London)*, v. A242, 1957.

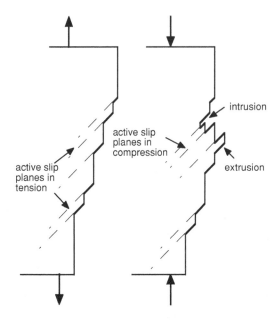

Figure 5.18. Sketch illustrating the formation of intrusions and extrusions by slip. From W. F. Hosford, *Mechanical Behavior of Materials*, Cambridge, 2005.

Figure 5.19. Clamshell markings on a fatigue fracture surface. The crack initiated at the left. The rough area near the right is the final fracture. Courtesy W. H. Durant.

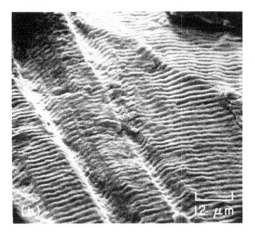

Figure 5.20. Striations on a fatigue fracture surface. Neighboring marks corresponds to the positions of the crack front on successive cycles. The distance between markings is the distance advanced by the crack in one cycle. From American Society for Metals, *Metals Handbook*, 8th ed., v. 9, 1974.

Toughness

The fracture behavior of a material is usually governed by the amount of energy it absorbs before breaking. The Charpy impact test is one way of characterizing a material's toughness. The energy to break a small, notched specimen (Figure 5.21) is measured. For steels and plastics, this energy is very temperature dependent. Figure 5.22 shows the temperature dependence of the energy for a typical steel. The tendency of a material to break in a brittle manner increases with three factors: lower temperatures, higher rates of loading, and the presence of notches.

Higher strength steels tend to have higher transition temperatures. They are more inclined to break in a brittle manner. Figure 5.23 shows the effect of carbon content on the transition temperature of plain carbon steels. Face-centered cubic metals including aluminum, copper-base alloys, and austenitic stainless steels are not embrittled by low temperatures. Temperature has only a minor effect on the toughness of most nonferrous metals.

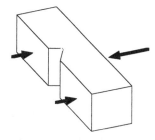

Figure 5.21. Charpy specimen with a V-notch.

Figure 5.22. Impact energy vs. temperature.

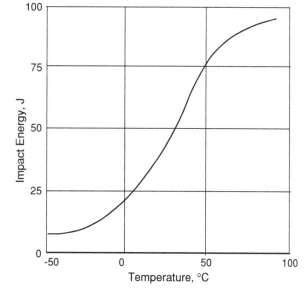

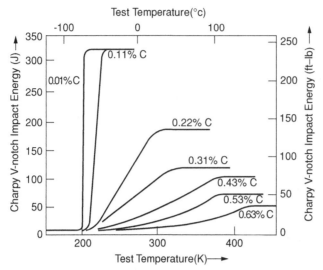

Figure 5.23. The Charpy ductile-brittle transition temperature increases with carbon content. From J. A. Rinebolt and W. J. Harris, *Trans ASM*, v. 43, 1951.

There is a sharp drop of toughness of a linear polymer below its glass transition temperature. Most ceramics are very brittle at room temperature. Their transition temperatures are very high.

Fracture Mechanics

Brittle materials are very sensitive to pre-existing cracks. The stress, σ_f, to cause fracture of a brittle material is given by

$$\sigma_f = K_{Ic}/[f\sqrt{\pi a}], \qquad\qquad 5.5$$

where K_{Ic} is the fracture toughness, a is the length of a crack extending to the surface, and f is a geometric factor and is usually a little greater than one for short cracks. For central cracks, a is half of the crack length. Figure 5.24 shows how f depends on geometry.

Example Problem 5–3:

A structural member of steel, 10 m long, 1 cm thick, and 10 cm wide, is under a stress tensile load of 625 kN. If the fracture toughness of the steel is 105 MPa$\sqrt{m}$, how long could a crack be before it fails catastrophically?

Solution: The stress is $\sigma = 625 \times 10^3/(10^{-2})(10^{-1}) = 625 \times 10^6$ Pa. Substituting into $\sigma_f = K_{IC}/\sqrt{(\pi a)}$, $a = (K_{IC}/\sigma_f)^2/\pi = (105/625)^2/\pi = 0.009$ m or 9 mm.

Usually K_{Ic} is inversely related to yield strength. Figure 5.25 shows how K_{IC} decreases in a 4340 steel as it is heat treated to have higher yield strengths.

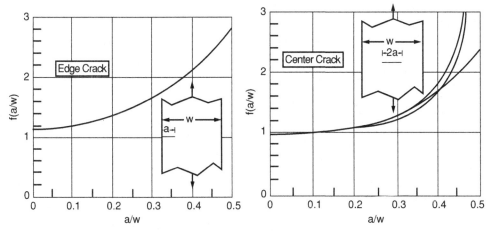

Figure 5.24. Dependence of the geometric factor, f, on specimen shape and crack length. From W. F. Hosford, *Mechanical Behavior of Materials*, Cambridge, 2005.

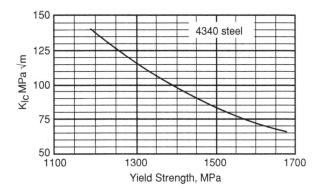

Figure 5.25. Decrease of fracture toughness with yield strength of a 4340 steel. From W. F. Hosford, *ibid*.

Thermal Shock

Brittle ceramics are likely to fracture when subjected to sudden temperature changes. If a hot ceramic is suddenly cooled, the tendency for the cooler outside to contract is resisted by the hot interior. This constraint causes tensile stresses on the outside, which may exceed the fracture strength.

If contraction is completely prevented in the surface,

$$\varepsilon = 0 = \alpha \Delta T + (1 - \upsilon)\sigma/E, \qquad 5.6$$

so the level of tensile stress is

$$\sigma = \alpha \Delta T E/(1 - \upsilon). \qquad 5.7$$

Thus the susceptibility to thermal shock increases with α and E. Table 5.1 shows the thermal expansion coefficients of several glasses. The sensitivity to thermal shock decreases in the same order that α decreases.

Table 5.1. *Thermal expansion coefficients, α, of several glasses*

Glass	Coefficient of thermal expansion $10^{-6}/K$
Soda–lime	9
Pyrex	2.5
Vycor	0.6
Silica	0.5

Example Problem 5–4:

The surface of a piece of soda-lime glass is suddenly cooled from 600 to 25 °C and restrained by the interior from contracting. Calculate the stress in the surface assuming that the thermal expansion coefficient is $9 \times 10^{-6}/K$, Young's modulus is 70 MPa, and Poisson's ratio is 0.30.

Solution: Substituting in $\sigma = \alpha \Delta TE/(1 - \upsilon)$, $\sigma = (9 \times 10^{-6}/K)(575 °C) \times (70 \times 10^{-6} Pa)/(1–0.3) = 517$ kPa.

Delayed Fracture

While glasses are not susceptible to fatigue, they may fail suddenly if loaded for long periods, particularly in wet environments. Such delayed fracture is not common in metals (except in cases of hydrogen embrittlement of steels), but sometimes does occur in polymers. It is often called *static fatigue*. The phenomenon in glass is sensitive to temperature and prior abrasion of the surface. Most importantly, it is very sensitive to environment. Cracking is much more rapid with exposure to water than if the glass is kept dry (Figure 5.26) because water breaks the Si-O-Si bonds by the reaction -Si-O-Si- + H_2O → Si-OH + HO-Si.

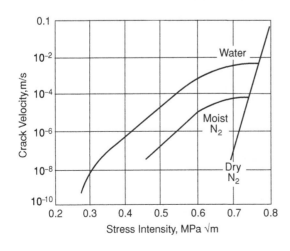

Figure 5.26. The effect of environment on crack velocity in a silicate glass. From *Engineered Materials Handbook, Vol. 4, Ceramics and Glasses*, ASM, 1991.

Figure 5.27. Brittle fracture of a ship. From C. F. Tipper, *The Brittle Fracture Story*, Cambridge, 1963.

Notes of Interest

The Great Boston Molasses Flood occurred on January 15, 1919. A large molasses storage tank suddenly burst, and a wave of molasses ran through the streets at an estimated 35 miles per hour, killing 21 and injuring 150. Residents claim that on hot summer days the area still smells of molasses. This was the first brittle fracture of a large structure. Other failures of large engineering structures include fractures in pipelines and ships.

Figure 5.27 shows a ship that failed in a harbor, and Figure 5.28 shows a pipeline that failed while it was being tested.

Figure 5.28. Pipeline that failed during testing. From E. R. Parker, *Brittle Behavior of Engineering Structures,* John Wiley and Sons, 1957.

Problems

1. Data for the steady-state creep of a carbon steel are plotted in Figure 5.29. Determine the activation energy, Q, in the equation $\dot{\varepsilon} = f(\sigma)\exp[-Q/(RT)]$.

2. A part made from aluminum alloy 7075-T4 is subjected to 200,000 cycles at 250 MPa and 40,000 cycles at 300 MPa. According to Miner's rule, how many additional cycles can it withstand at 200 MPa without failure? Refer to Figure 5-13.

3. Describe how you would expect an increase in temperature to change the *S-N* curve of an aluminum alloy.

4. Steels of higher yield strengths generally have lower toughnesses. Figure 5.25 shows how K_{Ic} depends on yield strength of 4340 steel after different heat treatments. Assume that inspection can detect any crack greater than 2 mm. What is the highest stress that can be applied without fear of either fracture or yielding? (Assume $f = 1.12$.)

5. A wing panel of a supersonic aircraft is made from a titanium alloy that has a yield strength of 1,035 MPa and toughness of $KIc = 55$ MPa$\sqrt{m}$. It is 3.0 mm thick, 2.40 m long, and 2.40 m wide. In service it is subjected to a cyclic stress of $\pm\,600$ MPa, which is not enough to cause yielding but does cause gradual crack growth of a pre-existing crack normal to the loading direction at the edge of the panel. Assume that f for fracture $= 1.12$, and the crack is initially 0.5 mm long and grows at a constant rate of $da/dN = 120\ \mu$m/cycle. How many cycles can the wing panel stand before fracture?

6. Frequently the *S-N* curves for steel can be approximated by a straight line between $N = 102$ and $N = 106$ cycles when the data are plotted on a log-log scale as shown in Figure 5.30 for SAE 4140 steel. This implies $S = AN^b$, where A and b are constants.

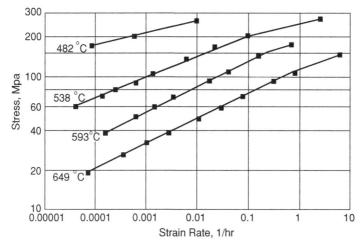

Figure 5.29. Creep rates of carbon steel at several temperatures. Data from P. N. Randall, *Proc. ASTM*, v. 57, 1957. From W. F. Hosford, *Mechanical Behavior of Materials*, Cambridge, 2005.

a. Find b for a certain part made from 4140 steel (Figure 5.30).
b. Fatigue failures occur after five years. By what factor would the cyclic stress amplitude have to be reduced to increase the life to ten years? Assume the number of cycles of importance is proportional to time of service.

7. Discuss whether increasing the strength of a material increases its resistance to fracture.

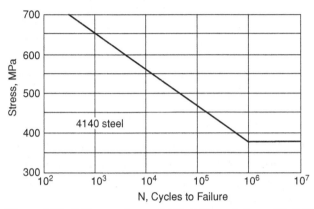

Figure 5.30. *S-N* curve for an SAE 4140 steel. From W. F. Hosford, *Mechanical Behavior of Materials*, Cambridge, 2005.

6 Annealing

Stages of Annealing

Annealing is the heating of a metal to soften it after it has been cold worked. Most of the energy expended in cold work is released as heat during the deformation; however, a small percentage is stored by lattice imperfections. There are three stages of annealing. In order of increasing time and temperature, they are:

1. Recovery, in which most of the electrical conductivity is restored. There is often a small drop in hardness; overall grain shape and orientation remain unchanged.
2. Recrystallization, which is the replacement of cold-worked grains by new ones with new orientations, a new grain size, and a new shape. Recrystallization causes the major hardness decrease.
3. Grain growth, which is the growth of recrystallized grains at the expense of other recrystallized grains.

Recovery

The energy release during recovery results from decreased numbers of point defects and rearrangement of dislocations. Most of the increase of electrical resisitivity during cold work is attributable to vacancies. These largely disappear during recovery so the electrical resistivity drops (Figure 6.1) before any major hardness changes occur. During recovery, residual stresses are relieved by creep, and this decreases the energy stored as elastic strains. Recovery causes no changes in microstructure that are observable under a light microscope.

Recrystallization

Recrystallization is the formation of new grains in cold-worked material. The new grains must first nucleate and then grow. Figure 6.2 shows the progress of recrystallization at $310\,°C$ of aluminum that had been cold worked 5%.

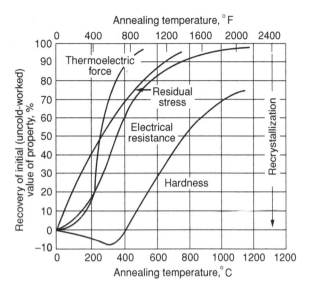

Figure 6.1. Property changes in tungsten during recovery. Note that the electrical conductivity improves before any major change in hardness. From A. Guy, *Elements of Physical Metallurgy*, 3rd ed., Addison-Wesley, 1958, p. 439.

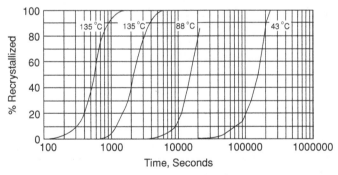

Figure 6.2. Progress of recrystallization at 310 °C of aluminum that had been cold worked 5%. From A. Guy, *ibid*.

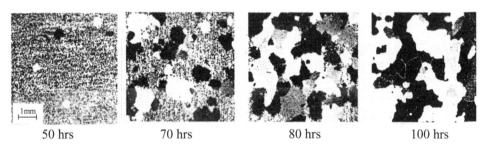

Figure 6.3. Isothermal recrystallization of 99.999% pure copper cold worked 98%. Data from B. F. Decker and D. Harker, *Trans. AIME*, v. 188, 1950.

Figure 6.3 shows how recrystallization depends on temperature and time for high-purity copper. The time-temperature dependence of recrystallization follows an Arrhenius relation,

$$t = A \exp(Q/RT), \hspace{3cm} 6.1$$

where t is the time to achieve structure with a certain fraction recrystallized, T is the absolute temperature, Q is an activation energy, A is a constant, and R is the gas constant.

Figure 6.4 is a plot of the time for 50% recrystallization of the copper in Figure 6.3 as a function of reciprocal temperature.

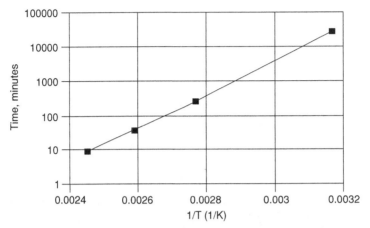

Figure 6.4. An Arrhenius plot of the data in Figure 6.3. The slope equals Q/R. From W. F. Hosford, *Physical Metallurgy*, CRC, 2005.

Example Problem 6–1:

Using Figure 6.3, determine the activation energy for recrystallization of copper.

Solution: Writing the Arrhenius equation as $t_2/t_1 = \exp[(Q/R)(1/T_2 - 1/T_1)]$ and solving for Q, $Q = R\ln(t_2/t_1)/(1/T_2 - 1/T_1)$. Taking points for 50% recrystallization, $t = 570$ s at $135\,°C$ and $t = 150,000$ at $43\,°C$, and substituting

$$Q = -8.32\ln(570/150,000)/(1/408 - 1/316) = 65,000 \text{ J/mole.}$$

Example Problem 6–2:

Estimate the time required to recrystallize this copper at $25\,°C$.

Solution: Substituting $T_3 = 298\,K$, $T_2 = 316\,K$, and $t_2 = 150,000$ s into $t_3/t_2 = \exp[(Q/R)(1/T_3 - 1/T_2)]$ and solving for t_3, $t_3 = 150,000 \exp[(65,000/8.23)(1/298 - 1/316)] = 668 \times 10^3$ s $= 185$ days.

It is common to define a recrystallization temperature as the temperature at which the microstructure will be 50% recrystallized in 30 minutes. As a rule of thumb, the recrystallization temperature for a metal of commercial purity is in the range of 1/3 to 1/2 of its absolute melting point (Figure 6.5); however, the recrystallization temperature is affected by purity, the amount of cold work, and the prior grain size. Solid solution impurities tend to segregate at grain boundaries and, therefore, slow

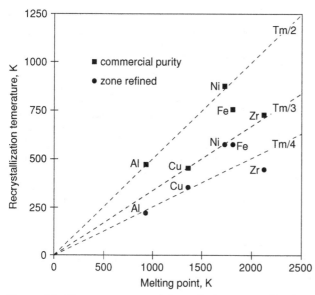

Figure 6.5. For most commercial metals the recrystallization temperature is between 1/3 and 1/2 of the absolute melting point. For zone refined metal, the ratio of recrystallization temperature to melting point is nearer 1/4. Data from O. Dimitrov et al. in *Recrystallization of Metallic Materials*, Riederer-Verlag GMBH, 1978. From W. F. Hosford, *Physical Metallurgy*, CRC, 2005.

their motion, thus raising the recrystallization temperature. Increased amounts of cold work and finer prior grain sizes increase the rates of nucleation of new grains and, therefore, lower recrystallization temperatures.

These same variables also affect the grain size that results from recrystallization. Increasing the amount of cold work increases both the nucleation rate and the growth rate, but the nucleation rate is increased more than the growth rate so increased cold work decreases the grain size that results from recrystallization. Figure 6.6 shows this effect. Below about 5% cold work, recrystallization did not occur.

A mathematical analysis of the kinetics of recrystallization by nucleation and growth is covered in Appendix 6.

In summary:

1. The recrystallization temperature goes down with increased purity, increased cold work, and finer prior grain size.
2. The recrystallized grain size decreases with increased cold work and finer prior grain size and increases with purity.

Recrystallization generally does not produce equiaxed grains, nor does it result in randomly oriented grains. Usually recrystallized grains are elongated in the direction of prior working because arrays of inclusions, strung out during working, slows growth in the thickness direction.

Mechanical deformation causes certain preferred orientation or crystallographic textures, which in turn cause some properties to be anisotropic. Recrystallization

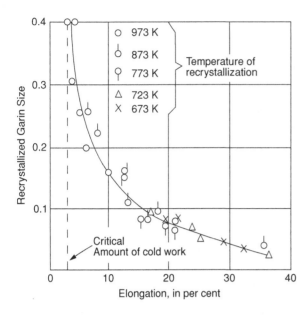

Figure 6.6. The effect of the amount of cold work on the recrystallized grain size of copper. Below about 3% cold work, recrystallization did not occur. Note that the temperature at which recrystallization did occur had little effect on the recrystallized grain size. From J. S. Smart and A. A. Smith, *Trans. AIME*, v. 152, 1943, p. 103.

may change the preferred orientations, but it does not produce randomly oriented grains.

Grain Growth

Grain growth is the increase of the average grain size of recrystallized grains with continued heating. The driving force is the decrease of total surface with larger grains. Grain boundaries tend to move toward their centers of curvature, while meeting other grain boundaries at 120°. In two dimensions, large grains tend to have more than six neighbors while small grains have fewer than six neighbors. Therefore, the boundaries of large grains are outwardly concave, and those of small grains are inwardly concave as shown in Figure 6.7. This causes the large grains to grow and the small ones to shrink and finally disappear.

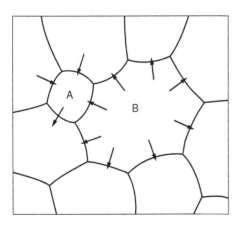

Figure 6.7. The large grain (B) has eight neighbors, so because grain boundaries meet at 120°, its boundaries are outwardly concave and tend to grow into neighbors. The small grain (A) has only four neighbors, so its boundaries are inwardly concave and it tends to shrink. From W. F. Hosford, *Physical Metallurgy*, CRC, 2005.

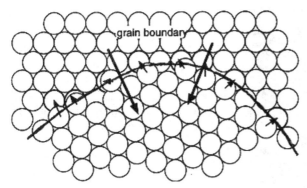

Figure 6.8. Atoms tend to move so as to have more correct near neighbors. This movement causes the grain boundary to move toward its center of curvature. From W. F. Hosford, *Physical Metallurgy*, CRC, 2005.

This can also be understood on an atomistic scale. Atoms on an outwardly curved grain boundary have fewer correct near neighbors than those on an inwardly curved boundary. As an atom moves to become part of the crystal with the concave boundary, it causes the boundary to move toward its center of curvature as indicated in Figure 6.8.

One might expect that the rate of grain growth would be proportional to the curvature, which would be proportional to $1/D$. In this case, the grain size would be given by

$$D^2 - D_o^2 = At, \qquad\qquad 6.2$$

where t is the time and A is a constant. However, because of second-phase inclusions and segregation of impurities to the boundaries, grain growth is better described by

$$D = kt^n, \qquad\qquad 6.3$$

where n is usually less than $1/2$.

Example Problem 6–3:

Calculate the decrease of surface energy per volume if the grain size of copper increases from 0.1 mm diameter to 1.0 mm diameter. The grain boundary surface energy of copper is the amount of $\gamma = 0.6\,\text{J/m}^2$.

Solution: Assuming a spherical shape for grains, the surface area per volume is $(4\pi r^2)/[(4/3\pi r^3) = 3/r$. Realizing that the surface is shared by two grains and $r = d/2$, surface energy per volume is $\gamma(3/d)$. The decrease of surface energy with growth of grains from 0.1 to 1.0 mm is $3\gamma(1/10^{-4} - 1/10^{-3}) = 27 \times 10^3\gamma = 16.2\,\text{kJ/m}^3$.

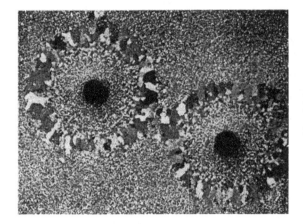

Figure 6.9. Recrystallized grain size near bullet holes in a tin sheet. From J. Czochralski in *Modern Physics*, Berlin.

Note of Interest

Figure 6.9 shows a sheet of tin through which two bullets were fired. It was etched a day or two afterward. Tin recrystallizes at room temperature. Note that the recrystallized grain size near the bullet holes is small. This is where there was the most deformation. Further from the bullet hole, the grain size is larger because the cold work was less. Further still, there was not enough deformation to cause recrystallization.

Problems

1. Using Figure 6.3, determine the recrystallization temperature of copper defined as the temperature at which the microstucture is 50% recrystallized in 30 minutes. Compare this with the rule of thumb that the recrystallization temperature for a metal of commercial purity is in the range of 1/3 to 1/2 of its absolute melting point, and explain any discrepancy.
2. Figure 6.10 shows the heat released during recrystallization of copper at 170 °C.
   a. Find the area under the curve in J/mole.
   b. If all of the energy went into heating the copper, the temperature would rise by how much? The specific heat of copper is 0.39 J/g °C.
3. Figure 6.11 shows how the hardness of cold-worked brass changes during annealing.
   a. After annealing at 300 °C, the lot cold rolled 40% is softer than the lot rolled 20%. Explain.
   b. After annealing both at 350 °C, the lot cold rolled 20% is softer than the lot rolled 40%. Explain.
4. Figure 6.12 shows the progression of recrystallization in Fe-3.5%Si after 60% cold reduction. Determine the activation energy for recrystallization.

Figure 6.10. Heat released from copper cold worked 40% during an anneal at 170 °C. Data from P. Gordon, *Trans AIME*, v. 203, 1955. Data for copper: AW = 63.54, density = 8.93 Mg/m³, heat capacity = 386 J/kg-K, atomic diameter = 0.256.

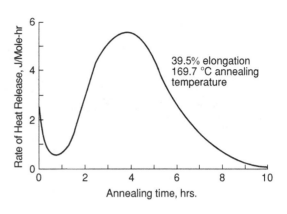

Figure 6.11. Hardness changes on annealing brass that has been cold worked different amounts. From W. F. Hosford, *Physical Metallurgy*, CRC, 2005.

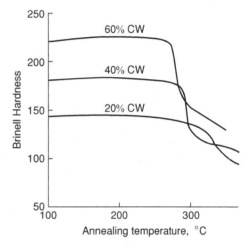

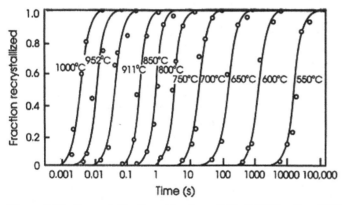

Figure 6.12. Recrystallization kinetics of Fe-3.5%Si after 60% cold work. From F. J. Humphries and M. Hatherly, *Recrystallization and Related Phenomena*, Pergamon, 1995. Data from G. R. Speich and R. M. Fisher, *Recrystallization, Grain Growth and Textures*, ASM, 1966.

7 Iron and Steel

Steels

Steels are iron-base alloys usually containing carbon. Figure 7.1 shows the iron-carbon phase diagram. Below 911 °C and between 1410 °C and the melting point, iron has a bcc crystal structure called *ferrite*. Between 1410 °C and 911 °C it has an fcc crystal structure called *austenite*. Austenite dissolves much more carbon interstitially than ferrite. On slow cooling below 727 °C, the austenite transforms by a *eutectoid reaction* into ferrite and iron carbide or *cementite* (which contains 6.7% C). The ferrite and cementite form alternating platelets called the *eutectoid* temperature. The resulting microstucture is called *pearlite* (see Figure 7.2).

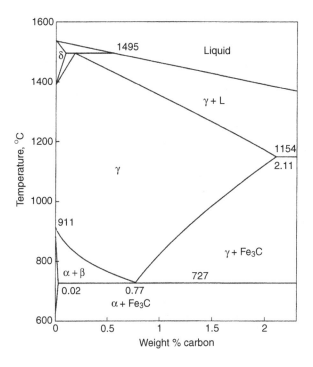

Figure 7.1. The iron-carbon phase diagram. From W. F. Hosford, *Physical Metallurgy*, CRC, 2005.

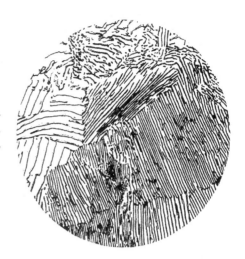

Figure 7.2. Microstructure of pearlite. Note the alternating platelets of ferrite (white) and iron carbide (dark). From *Making, Shaping and Treating Steels*, 9th ed., United States Steel Corp., 1971.

Pearlite Formation

When a steel containing less than 0.77% C (*hypo-eutectoid steel*) is slowly cooled, some ferrite forms before any pearlite. A steel containing more than 0.77% C (*hyper-eutectoid steel*) will form some cementite before any pearlite. The formation of pearlite from austenite on cooling requires diffusion of carbon from ahead of the advancing ferrite platelets to the advancing carbide platelets as indicated in Figure 7.3. Because diffusion takes time, pearlite formation is not instantaneous. The rate at which pearlite forms depends on how much the temperature is below 717 °C. The rate of diffusion increases with temperature, but the driving force for the transformation increases as the temperature is lowered. The result is that the rate of transformation is fastest between 500 and 600 °C, as indicated schematically in Figure 7.4.

Example Problem 7–1:

Calculate the fraction of proeutectoid cementite in a steel containing 1.10% C slowly cooled through the eutectoid temperature.

Figure 7.3. The formation of pearlite requires the diffusion of carbon to the growing cementite platelets.

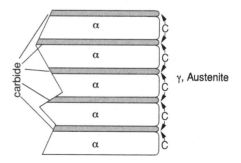

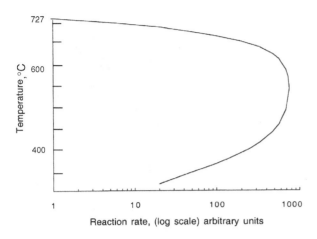

Figure 7.4. The rate of pearlite formation reaches a maximum at about $600\,°C$.

Solution: Using the lever law just above the eutectoid temperature, $f_{cem} = (1.00 - 0.80)/(6.67 - 0.80) = 3.4\%$.

Alloying Elements in Steel

Alloying elements affect both the eutectoid temperature and the amount of carbon at the eutectoid, as shown in Figure 7.5. Manganese and nickel are more soluble in austenite than ferrite, and therefore lower the eutectoid temperature. Silicon and elements with a bcc crystal structure have a greater solubility in ferrite, and therefore

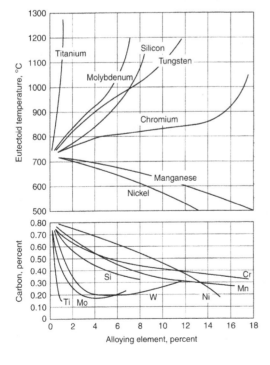

Figure 7.5. Effect of alloying elements on the eutectoid temperature of steels (top) and the eutectoid composition (bottom). From *Making, Shaping and Treating Steels*, 9th ed., United States Steel Corp., 1971.

raise the eutectoid composition. All common alloying elements lower the carbon content of the eutectoid.

Example Problem 7–2:

Estimate the eutectoid temperature and composition of a 4340 steel containing 1.65% Ni, 0.70% Cr, 0.25% Mo, 0.45% Mn, 0.25% Si, and 0.40% C.

Solution: Referring to Figure 7.5, the eutectoid composition is decreased $0.04/2 =$ 0.02 per % Ni, 0.1 per % Cr, 0.25 per % Mo, 0.06 per % Mn, and 0.13 per % Si so the eutectoid composition is $0.78 - (0.02)(1.65) - (0.1)(0.70) - (0.25)(0.25) - (0.06)(0.45) - (0.13)(0.25) = 0.655$.

The eutectoid temperature is decreased about $4°$ per % Mn and $7°$ per % Ni, and raised about $20°$ per % Cr and $30°$ per % Si and $60°$ per % Mo, so the change of eutectoid temperature is $-7(1.65) - 4(0.45) + 20(0.65) + 30(0.25) + 60(0.25) = +22°C$.

Martensite Formation

Figure 7.6 is an isothermal transformation diagram showing the time required for transformation of austenite to ferrite and carbide as a function of temperature. If the cooling is fast enough there may not be time for pearlite to form. In this case, the austenite becomes so unstable that a new phase, *martensite*, forms by a shearing of the lattice. Martensite has a body-centered tetragonal crystal structure and can be thought of as ferrite supersaturated with carbon in interstitial solid solution. Quenching to avoid prior transformation to pearlite and thereby allowing austenite to transform to martensite is the basis of hardening steels.

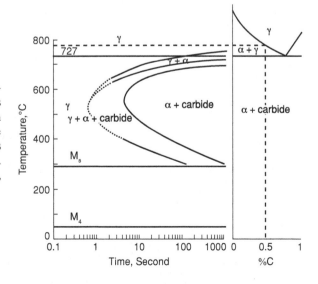

Figure 7.6. An isothermal transformation diagram for a steel containing less than 0.77% C. Some ferrite must form before pearlite. Below about 300°C, the austenite is so unstable that it transforms by shear into a new phase, martensite. From W. F. Hosford, *Physical Metallurgy*, CRC, 2005.

Nature of Martensite

The hardness and tetragonality of martensite depend only on carbon content. Figure 7.7 shows that the tetragonality increases with carbon content. The fact the lattice parameters extrapolate to that of ferrite reinforces the idea that martensite can be thought of as ferrite supersaturated with carbon. Figure 7.8 shows that the hardness of martensite increases with carbon content.

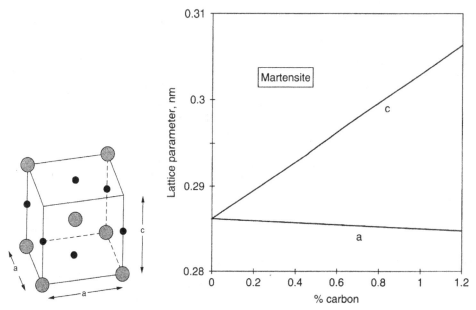

Figure 7.7. The martensite unit cell (left). The ratio of the c lattice parameter to the a parameter increases with carbon content (right). From W. F. Hosford, *Physical Metallurgy*, CRC, 2005.

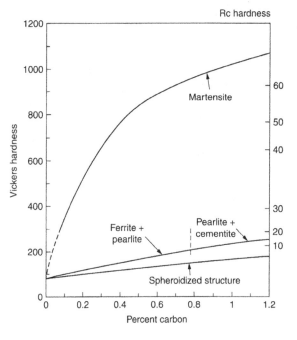

Figure 7.8. The hardness of martensite increases with carbon content. Note that the hardness extrapolates to the hardness of ferrite at 0% C. The hardnesses of pearlitic and spheroidized structures are shown for comparison. From W. F. Hosford, *Physical Metallurgy*, CRC, 2005.

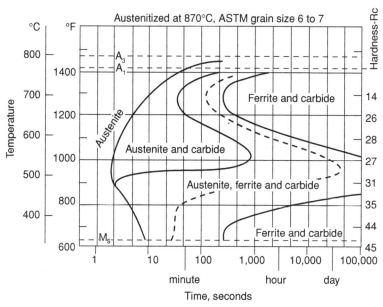

Figure 7.9. Isothermal transformation diagram for a 4340 steel containing 0.42% C, 0.78% Mn, 1.79% Ni, 0.80% Cr, and 0.33% Mo. Note that the time for transformation is much longer than for the eutectoid steel in Figure 7.5. From *Making, Shaping and Treating Steels*, 9th ed., United States Steel Corp., 1971.

Hardenability

Alloying elements slow the transformation of austenite to pearlite because they must diffuse to either the growing cementite or ferrite platelets. Nickel and silicon are not carbide-forming elements and therefore segregate to the ferrite. Manganese, chromium, molybdenum, and tungsten are carbide formers so they segregate to the cementite. In either case, the diffusion of these elements in substitutional solid solution is very much slower than the diffusion of carbon, so they slow the transformation of austenite to pearlite.

If parts made from plain carbon steels are too thick, their centers may not cool fast enough to avoid pearlite formation. We say they have insufficient *hardenability*. Steels containing alloying elements will harden under somewhat slower cooling rates. Figure 7.9 shows the isothermal transformation diagram of 4340 steel.

The Jominy end-quench test is a simple way to assess the hardenability of a steel. Water is sprayed on the end of a bar that has been heated to form austenite, as shown in Figure 7.10. This creates a gradient of cooling rates along the bar, with the quenched end cooling very rapidly. Hardness measurements are then made along the length of the bar. Figures 7.11, 7.12, and 7.13 show typical results.

Figures 7.11 through 7.13 illustrate three points. First, the hardness at the quenched end depends on the carbon content and is independent of alloy content. Second, the depth of hardening (hardenability) increases with alloy and carbon contents. Finally, hardenability increases with austenite grain size. This is because pearlite nucleates on grain boundaries, so there are fewer nucleation sites with a coarse grain size.

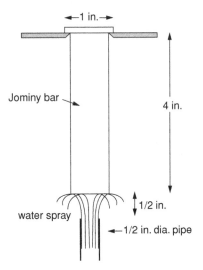

←1 in.→

Jominy bar

4 in.

1/2 in.

water spray

←1/2 in. dia. pipe

Figure 7.10. Jominy end-quench test. From W. F. Hosford, *Physical Metallurgy*, CRC, 2005.

Example Problem 7–3:

When a test part made from a 1060 steel with an ASTM grain size of 2 was quenched in stirred oil, the hardness at a critical place was Rc 30. What would the hardness at that spot be on a 4140 steel given the same quench?

Solution: According to Figure 7.11 a hardness of Rc 40 in the 1060 steel corresponds to the cooling rate at 15 mm. According to Figure 7.11 the hardness of a 4140 steel at that Jominy distance would be Rc 51.

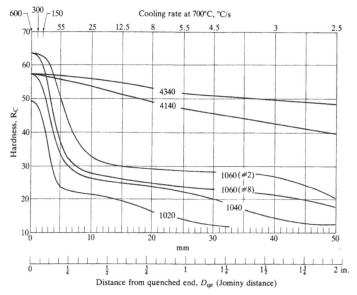

Figure 7.11. Jominy end-quench curves for several steels. From L. H. Van Vlack, *Elements of Materials Science*, 5th ed. Addison-Wesley, 1985, p. 472.

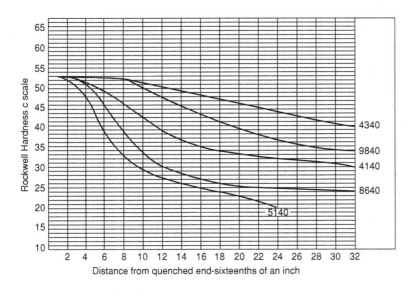

	C	Mn	Ni	Cr	Mo
4340	0.40	0.70	1.65	0.70	0.20
9840	0.40				
4140	0.40	0.75	—	0.80	0.15
4140	0.40	0.75	0.40	0.80	0.15
8640	0.40	0.75	—	0.40	0.15
5140	0.40	0.70	—	0.70	—

Figure 7.12. Jominy end-quench curves for steels of varying alloy content. All contain 0.40% C. Note that the depth of hardening increases with alloy content. From *Making, Shaping and Treating Steels*, 9th ed., United States Steel Corp., 1971.

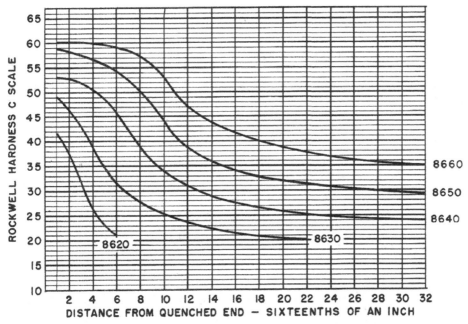

Figure 7.13. Jominy end-quench curves for steels of varying carbon content. All contain 0.7 to 1.0% Mn, 0.4 to 0.7% Ni, 0.4 to 0.5% Cr, and 0.15 to 0.25% Mo. Note that the depth of hardening increases with carbon content. From *Making, Shaping and Treating Steels*, 9th ed., United States Steel Corp., 1971.

Table 7.1. *Relative effectiveness of alloying elements on increasing hardenability per weight percent*

Ni	0.36
Si	0.70
Cr	2.16
Mo	3.0
Mn	3.33

The relative effectiveness of various alloying elements at low concentrations in increasing hardenability is given in Table 7.1.

Quenching

The rate at which a given spot on a part cools depends on its location (e.g., surface, center), the size of the part, and the quenching medium. Various quenching media are used, including iced brine, water, oil, and molten salts. The reason that quenching media other than water are not often used is that a part may crack during quenching because of high temperature gradients. Temperature gradients may also cause residual stresses and distortion of shape. Figure 7.14 shows how the cooling rates at different locations in a 2-in. diameter bar depend on quenching media. Some tool steels with very high alloy content harden on air cooling.

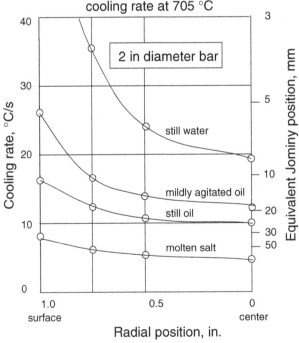

Figure 7.14. Cooling rates (at 705 °C) at various positions in a 2-in. diameter bar for different quench media. Note that the variation of cooling rate with position decreases with milder quenches.

Tempering

Although martensite can be very hard, it is too brittle for most applications. Therefore it is almost always reheated to temper it. Tempering at low temperatures relieves residual stresses and causes softening. At somewhat higher temperatures, tempering causes the martensite to break down. The carbon content of the martensite decreases as transition carbides form. Details of the tempering reactions are beyond the scope of this book, but increased tempering time and temperature cause continuous decrease of hardness. Figure 7.15 shows how the hardness of plain carbon steel decreases with tempering temperature. The Arrhenius equation can be applied to tempering. For the same hardness, the tempering times and temperatures are given by

$$t = A\exp(+Q/RT), \qquad\qquad 7.1$$

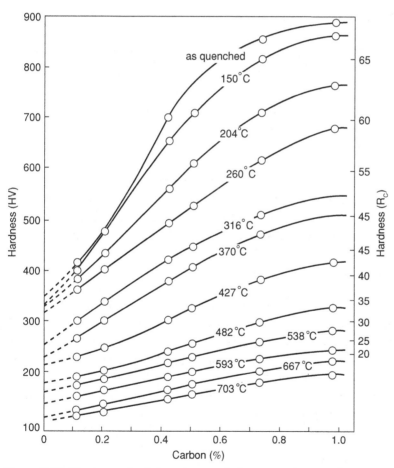

Figure 7.15. Hardness of a plain carbon steel after tempering at various temperatures for one-half hour. Adapted from R. A. Grange, C. R. Hribal, and L. F. Porter, *Met Trans A*, v. 8A, 1977. From W. F. Hosford, *Physical Metallurgy*, CRC, 2005.

where A is a constant. The hardness can be expressed as

$$H = f[t \exp(-Q/RT)], \tag{7.2}$$

Example Problem 7–5:

Figure 7.16 shows the tempering characteristics of a 1080 steel. Determine the activation energy for tempering to a hardness of Rc 45.

Solution: $Q = R \ln(t_2/t_1)/(1/T_2 - 1/T_1)$. Substituting $t_2 = 100{,}00$ s at $300\,°C = 573$ K and $t_1 = 250$ s at $400\,°C = 673$ K, $Q = 8.23\ln(100{,}000/250)/(1/573 - 1/673) = 190$ kJ/mole.

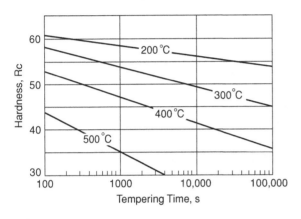

Figure 7.16. Tempering characteristics of a 1080 steel. From W. F. Hosford, *Physical Metallurgy*, CRC, 2005.

Different amounts of tempering are appropriate for different applications. High hardness is desirable for maintaining sharp cutting edges, and high toughness is needed where impact loading is possible. Razor blades and knife blades are only slightly tempered. More tempering is used for chisels. Hammers are tempered even more.

Carbide-forming alloying elements such as Cr, V, Mo, and W slow tempering because they must diffuse from the martensite to the growing carbides. Steels containing large amounts of carbide formers may undergo *secondary hardening* after high-temperature tempering. This results from the formation of a fine dispersion of very strong carbides. Such steels are used where hardness must be maintained at high temperatures, such as in tools for high-speed machining. The effects of molybdenum and chromium are shown in Figure 7.17.

Advertisements that use the word *tempering* often imply hardening, whereas tempering almost always softens steel.

Special Heat Treatments

Normalizing involves heating a steel into the austenitic temperature range and then air cooling. It is used primarily on plain carbon steels and results in a structure of

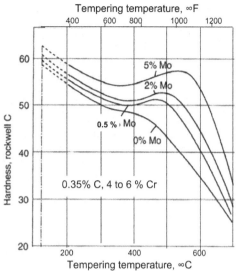

Figure 7.17. Secondary hardening in steels containing chromium and molybdenum. From *Making, Shaping and Treating Steels*, 9th ed., United States Steel Corp., 1971.

ferrite and pearlite (or in the case of steels with more than 0.77% carbon, primary cementite and pearlite).

Austempering involves quenching from the austenite temperature range into a molten salt bath held just above the Ms temperature and holding until transformation is complete. The product is called bainite, which is a very fine dispersion of carbide in ferrite. It is useful in many applications because of its hardness.

Carburizing is a process that develops very hard surfaces for wear resistance. Carbon is diffused into the surface of a part from a gas containing CO. The part is then quenched and tempered. *Nitriding* can be used to form very hard surfaces on steel that contains aluminum. The source of nitrogen is dissociated ammonia. The hardness results from the formation of aluminum nitride. The depth of the nitrided layer is typically much thinner than carburized layers.

When a conductive metal is placed in a varying magnetic field, eddy currents are induced that lead to heating. The depth to which the field penetrates depends on the frequency. At very high frequencies, the depth of penetration is *very* small. Because of this, the surfaces of a steel part can be heated into the austenitic range. As soon as the field is removed, the cool interior quenches the surface fast enough to form martensite. Passing the part through a second induction coil can be used to temper the martensite. This process, called *induction hardening*, produces a hard surface without the necessity of any chemical change.

Flames and lasers can be used to heat and harden only the surface of parts.

Steel Classification System

The AISI-SAE system for classifying higher carbon and alloy steels is covered in Table 7.2. The last two digits indicate the carbon content in hundredths of a percent to the nearest 0.05%. (Three digits are used for steels with 1% carbon or more.) The

Table 7.2. *AISI and SAE classes of steels*

Number	Composition
10xx	plain carbon (with about 0.50% Mn)
11xx	plain carbon resulfurized for machinability
15xx	1.0 to 2.0% Mn
31xx	1.25% Ni, 0.65% Cr
40xx	0.2 to 0.30% Mo
41xx	0.4 to 1.2% Cr and 0.08 to 0.325% Mo
43xx	1.65 to 2.0% Ni, 0.4 to 0.90% Cr, 0.20 to 0.30% Mo
51xx	0.70 to 1.10% Cr

first two digits indicate the main alloying elements. For example, a 4340 steel contains 0.38 to 0.43% C, 0.60 to 0.80% Mn, 0.20 to 0.35% Si, 1.65 to 2.00% Ni, 0.40 to 0.60% Cr, 0.20 to 0.30% Mo with < 0.035% P and < 0.040% S.

Low-carbon Steels

By far the largest tonnage of steels contain 0.10% carbon or less with no alloying elements except some manganese and silicon. These steels are classified as either *hot rolled* or *cold rolled*. Hot-rolled grades are somewhat cheaper because they are finished hot but have rough surfaces. They are used where surface appearance is not important, such as for I-beams and structural plate.

Cold-rolled steels are hot rolled, pickled in acid to remove scale, cold rolled, and finally annealed in an inert atmosphere. Auto bodies, appliances, and other consumer items are made from cold-rolled steel because of their very smooth surfaces. The defomation during stamping is often severe, so a major concern of these steels is formability. For that reason, the carbon content is usually less than 0.06%, and in some cases it is as low as 0.002%. In recent decades, concern for weight saving has led to the introduction of steels with somewhat higher yield strengths. *High-strength low-alloy steels* are strengthened by a combination of precipitation of carbonitrides, finer grain sizes, and solid solution effects. The formability of these steels is somewhat less, but adequate for most automobile parts. Dual-phase steels that are quenched from the ferrite-austenite region contain mostly ferrite with some martensite.

Example Problem 7–4:

A low-alloy steel containing 0.15% C is quenched from 750°C. Using the phase diagram and Figure 7.8, estimate the as-quenched hardness. At 750°C austenite contains 0.6% C and ferrite contains 0.02% C.

Solution: Using the lever law, the fraction of austenite before quenching would be $f_\gamma = (0.15–0.02)/(0.6 −0.02) = 22.4\%$. On quenching this would form martensite containing 0.6% C with a Vickers hardness of 825. The ferrite would have a Vickers hardness of about 90. The overall hardness should be the weighted average, $(0.224)(825) + (.776)(90) = 255$.

Stainless Steels

Alloying elements have very little effect on the corrosion behavior of steels. The exception to this is steels in which the chromium content is at least 12%. Such *stainless steels* are very resistant to corrosion in the presence of oxygen. There are three main groups of stainless steels. Ferritic stainless steels contain 12 to 18% chromium and very little carbon (generally less than 0.03%). Even when heated to temperatures just below their melting points, they maintain a bcc crystal structure. Austenitic stainless steels contain enough nickel so that they have an fcc crystal structure at all temperatures below melting. A typical composition is 18% Cr, 8% Ni, with very low carbon. They are more expensive than ferritic stainless steels and are non-magnetic. The third main class of stainless steels are martensitic. Martensitic stainless steels have compositions similar to ferritic stainless steels, but with enough carbon so that when heated they transform to austenite. The hardenability is so high that they form martensite even with slow cooling. Applications include razor and knife blades.

Cast Iron

Cast iron contains far more carbon than steel. Typical carbon contents range from 2 to 4% carbon with 1 to 3% silicon. With so much carbon and silicon, graphite tends to form on freezing, rather than cementite. With normal cooling in sand molds, graphite forms as flakes. Figure 7.18 shows a typical microstructure of a *gray cast iron*. Ductility is limited to a few percent elongation because of easy fracture along the graphite flakes. The name *gray cast iron* comes from the appearance of the fracture surfaces, which are almost entirely graphite.

Figure 7.18. Microstructure of gray cast iron. The black flakes are graphite. The white areas are ferrite and the gray areas are pearlite. From American Society for Metals, *Metals Handbook*, 8th ed., v. 8, 1982.

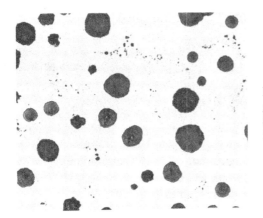

Figure 7.19. Ductile cast iron. Black spheroids are graphite. White areas are ferrite. From American Society for Metals, *Metals Handbook*, 8th ed., v. 7, 1972.

If the sulfur content is kept very low and magnesium or cerium is added to the melt just before casting, the graphite will form spheroids. The resulting product is called *ductile cast iron*. Figure 7.19 shows a typical microstructure. The mechanical behavior of ductile cast iron is similar to that of steel.

Both gray and ductile cast irons may be reheated to form a structure of graphite and austenite. The rate of cooling then determines whether the matrix is pearlite, ferrite, or a mixture of pearlite and ferrite. Rapid cooling can be used to form martensite. Austempering of ductile cast iron produces many useful products. Gray and ductile cast irons are cheaper than cast steel because they melt at much lower temperatures than steels, and they are easier to cast because they shrink very little on freezing.

White cast iron is a third grade. The carbon and silicon contents of white cast iron are lower than in gray and ductile irons, and the cooling rates are greater. These factors tend to prevent graphite formation so carbon occurs as iron carbide (cementite). White cast iron is very hard and brittle, and finds application only where the surface must be hard to resist wear.

Notes of Interest

Damascus steel was used in the Middle East for sword making from about 1100 to 1700 A.D. Damascus steel swords were renown for their sharpness and strength. The exact method used to make these swords is still in question. Their surfaces had an obvious pattern, which European sword makers tried to duplicate by various means. In the era of Damascus steel, steels that were hard, but brittle, could be produced by adding up to 2% carbon, or soft and malleable with about 0.5% carbon. For a good sword, steel should be both hard and tough so that it would hold an edge once sharpened, but not break when used in combat.

Metal smiths in India, perhaps as early as 100 B.C., developed a product known as wootz steel with a high carbon content. This steel contained a glassy phase that acted as a flux by reacting with impurities and allowing them to rise to the surface. This left a pure steel. This process was further refined in the Middle East, either using locally produced steels or by reworking wootz purchased from India. The exact process remains unknown, but carbides formed as bands within the blade. The carbides, which are much harder than the surrounding ferrite, permitted edges that would cut hard materials. The bands of softer ferrite gave the required toughness.

The first mass production of inexpensive steel from pig iron was by the Bessemer process. Henry Bessemer patented the process in 1855, although William Kelley had independently invented it in 1851. Before this, the only steel available was made in crucibles and was far too expensive for widespread use. The Bessemer process removed silicon and carbon from pig iron by blowing air into it through holes in the bottom of the vessel. The reactions generated enough heat to keep the metal molten.

Siemens developed and patented a regenerative furnace in the 1850s that could save fuel costs by recovering most of the heat. In 1865, Emile and Pierre Martin applied the regenerative furnace concept to making steel by the open-hearth process, which was slower than the Bessemer process but created larger heat and was easier to control.

Basic oxygen steelmaking was first used in 1952 in Linz, Austria. Often known as the LD process, it has replaced both Bessemer and open-hearth steelmaking. It is similar to the Bessemer process except that the vessels are much larger and the steel contains a significant fraction of steel scrap as well as molten pig iron. By 1992, the basic oxygen process had completely replaced Bessemer and open-hearth steelmaking in the United States.

Electric arc furnaces were first invented by Paul Héroult in France. Initially electric arc furnaces were used only for alloy steels. However, after World War II, their low capital costs allowed small mills to compete with large steelmakers. The charge is entirely scrap, so blast furnaces are not needed to make pig iron. Today they supply a major fraction of the entire steel production.

Problems

1. Calculate the fraction of pearlite that is cementite.
2. Describe how your answer would change for a steel containing both chromium and manganese as alloying elements.
3. A certain spot on a part of an 8640 steel was found to have a hardness of Rc 35 after quenching in oil. If a similar part of a 4340 steel were quenched the same way, what would the hardness of that spot be?
4. Alloying elements in steel slows the tempering process. Speculate about the reason for this.
5. Estimate the volume fraction of graphite in a ferritic gray cast iron containing 3.5% carbon. For simplicity, ignore the presence of silicon.

6. Why are greater amounts of alloying elements than in 4340 not commonly used in steels?

7. In a process called *marquenching*, austenite is quenched into a bath of molten salt and held long enough to eliminate thermal gradients – but not so long as to form bainite. Slow cooling then allows martensite to form. What is the advantage of this process over direct quenching to room temperature?

8 Nonferrous Metals

Aluminum

Aluminum and aluminum alloys are the most important nonferrous metals. Aluminum's density, ρ, is 2.7, which is about 1/3 of the density of iron. Its Young's modulus is 70 GPa (10×106 psi) which is also about 1/3 that of iron. The unique properties of aluminum that account for most of its usage are:

1. Good corrosion and oxidation resistance.
2. Good electrical and thermal conductivities.
3. Low density.
4. High reflectivity.
5. High ductility and reasonably high strength.

The uses of aluminum include foil, die castings, beverage cans, cooking and food processing, boats and canoes, and aircraft and automobile parts including sheet, engine blocks, and wheels. Aluminum's high reflectivity accounts for its use as foil for insulation and as reflective coatings on glass. Aluminum is used for power transmission lines and some wiring because of its high electrical conductivity. On an equal weight of cross section and equal cost bases, it is a better conductor than copper. Its high thermal conductivity is advantageous in its applications for radiators, air-cooled engines, and cooking utensils. The low density is important for lawn furniture, hand-held tools, and in cars, trucks, and aircraft. Aluminum's good strength and ductility is important in all structural uses where wrought products are used. Its chemical reactivity is important principally in its use in photoflash bulbs and the thermite reaction ($Al + Fe_2O_3 \rightarrow Fe + Al_2O_3$). Its corrosion and oxidation resistance are important in packaging (foil, cans), architectural applications, and watercraft.

Example Problem 8–1:

Calculate the ratio of wires of aluminum and copper having equal weight and equal length.

	Resistivity ohm · m	Density (kg/m^3)
Aluminum	29.9	2.7
Copper	17.0	8.9

Solution: $R_{Al}/R_{Cu} = (\rho_{Al}L_{Al}A_{Al})/(\rho_{Cu}L_{Cu}A_{Cu}) = (\rho_{Al}/\rho_{Cu})(A_{Cu}/A_{Al})$

For equal weight, $(d_{Al}L_{Al}A_{Al})/(d_{Cu}L_{Cu}A_{Cu})$, so $(A_{Cu}/A_{Al}) = (d_{Al}L_{Al})/(d_{Cu}L_{Cu}) = d_{Al}/d_{Cu}$.

Substituting, $R_{Al}/R_{Cu} = (\rho_{Al}d_{Al})/(\rho_{Cu}d_{Cu}) = (29.9 \times 2.7)/(17.0 \times 8.9) = 0.53$. Thus aluminum is a better conductor on an equal weight basis.

Aluminum is alloyed with manganese, magnesium, zinc, copper, and silicon. Manganese and magnesium are used for solid solution strengthening. Silicon is limited to alloys that are to be used in the cast form. Silicon makes the casting process simpler by lowering the melting point and reducing the liquid-to-solid shrinkage. The presence of a silicon phase, however, severely lowers the ductility so the silicon is kept low in alloys that are to be shaped mechanically. Alloying additions of copper and zinc form alloys that can hardened by heat treatment. Figure 8.1 shows the aluminum-rich end of the aluminum-copper phase diagram.

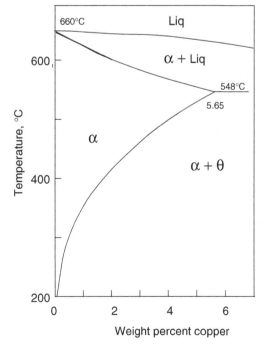

Figure 8.1. The aluminum-rich end of the aluminum-copper phase diagram.

If an alloy containing 4% copper is heated to 540 °C, all of the copper will dissolve in the aluminum-rich phase. This solution can be maintained by quenching rapidly to room temperature. On reheating the alloy to 190 °C, very fine precipitates form that harden the alloy, as indicated by Figure 8.2. This hardening mechanism is called *precipitation* or *age* hardening. Aluminum alloys containing zinc as well as those that contain magnesium and silicon in approximately the proportion to precipitate Mg_2Si are age-hardenable. Note that at long aging times the hardening is lost. This is because the average particle size increases by solution of the smaller particles and growth of the larger ones. This is known as *over-aging*. After very long times the hardness may drop below that of the quenched material because the solution hardening is lost as is the precipitation effect. The possibility of precipitation hardening also occurs in other alloy systems.

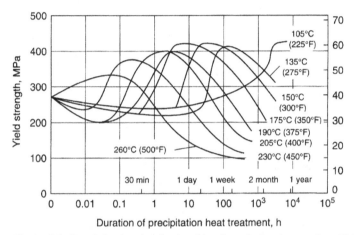

Figure 8.2. Precipitation hardening kinetics of aluminum alloy 2014. From *Aluminum and Aluminum Alloys*, ASM Int., 1993.

Example Problem 8–2:

Determine the activation energy for precipitation hardening of aluminum alloy 2014 (Figure 8.2) peak hardness from the times and temperatures for peak hardness.

Solution: Substituting $t_2 = 180$ hrs at $T_2 = 150\,°C = 423$ K and $t_1 = 0.05$ hrs at $T_2 = 260\,°C = 533$ K into $Q = R\ln(t_2/t_1)/(1/T_2 - 1/T_1)$, $Q = 140$ kJ/mole.

Example Problem 8–3:

Referring to Figure 8.2, the difference in the as-quenched and over-aged hardnesses reflects the loss of solid solution hardening. How much of the yield strength of the alloy aged one day at 175 °C can be attributed to the effect of the precipitates?

Solution: Assume that all of the copper has precipitated so there is no solution hardening after one day at 175 °C. The difference between yield strengths of

420 MPa for the precipitation hardened alloy and of 100 MPa for the overaged alloy without any solid solution strengthening, 320 MPa can be attributed to the effect of the precipitates.

Table 8.1 shows the designation system used for wrought aluminum alloys. The first number indicates the most important alloying element. The three numbers following this indicate different compositions. The letters after the numbers indicating composition give the heat-treated or cold-worked condition.

The numbers 4xxx were originally set aside for alloys with silicon as the major alloying element, but because of low ductility, there are no commercial alloys of this type.

Table 8.1. *Temper designations*

F	As fabricated
O	Annealed
H	Strain hardened
H-1x	Strain hardened only
H-2x	Strain hardened and partially annealed
H-3x	Strain hardened and stabilized
W	Solution treated (used only with alloys that will age at room temperature)
T	Heat treated to produce a stable temper (other than O, F, H, and W)
T-3	Solution treated, strain hardened, and naturally aged
T-4	Solution treated and naturally aged
T-5	Artificially aged only (after fast cooling from hot working)
T-6	Solution treated and artificially naturally aged

Number	Principal	Examples	Uses
	Alloy		
EC	99.45 +% Al		electrical grade
1xxx	commercially pure	1100 (0.12% Cu) 1060 (99.6% Al)	architectural, cookware
2xxx	copper	2014 (4.5 Cu, 0.5 Mg, 0.8 Mn, 0.8 Si)	structural forgings
		2024 (4.5 Cu, 1.5 Mg, 0.6 Mn)	aircraft, hardware
3xxx	manganese	3003 (1.2 Mn)	food and chemical processing
		3004 (1.2 Mn, 1 Mg, 0.2 Fe)	beverage cans, roofing
5xxx	magnesium	5052 (2.5 Mg, 0.25 Cr)	boats, trucks, busses
6xxx	Mg + Si	6061(1 Mg, 0.5 Si, 0.3 Cu)	trucks, furniture, boats
		6063 (0.7 Mg, 0.4 Si)	extrusions, architectural, irrigation
7xxx	Zn	7075(5.6 Zn, 1.6 Cu, 2.5 Mg)	aircraft
8xxx	misc. (including Li)		

The second digit (1 through 9) refers to the degree of hardening, where 8 corresponds to the hardness achieved by a 75% reduction. A third digit may be added to identify the surface condition.

For example, the designation 2014-T6 is for an alloy containing copper, which has been solution treated and allowed to naturally age harden at room temperature.

Table 8.2 lists a few common casting alloys. Most of the casting alloys contain major amounts of silicon. Aluminum alloys containing silicon undergo less shrinkage on freezing than most alloys. With 18% silicon, there is no shrinkage. This is because most of the silicon is elemental silicon with a diamond cubic crystal structure, which is less densely packed than the liquid.

Table 8.2. *Some aluminum casting alloys*

Number	Composition	Uses
296	4.5 Cu	aircraft fittings, pumps
356	7 Si, 3.5 Cu	engine blocks, transmissions, wheels
380	8 Si, 3.5 Cu	die castings
390	17 Si, 4,5 Cu,1 Fe, 0.5 Mg	die castings

Figure 8.3 is the aluminum-silicon phase diagram.

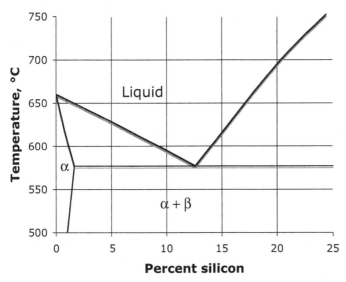

Figure 8.3. The aluminum-silicon phase diagram.

Copper and Copper Alloys

Pure copper is an excellent conductor of electricity and heat. Table 8.3 compares the electrical and thermal conductivities of several common metals. Where high conductivity is required, the oxygen-free high conductivity grade is used. All alloying elements lower the conductivity of copper. Copper's high electrical conductivity accounts for its use in wiring, and its high thermal conductivity accounts for its use in radiators and heat exchangers.

Table 8.3. *Conductivities of several metals*

Metal	Electrical conductivity $(n\Omega m)^{-1}$	Thermal conductivity $(Wm^{-1}K^{-1})$
Aluminum	37.7	223
Copper	59.77	395
Iron	10.3	75.6
Lead	4.84	35
Magnesium	22.5	154
Silver	69	420

Copper is also quite corrosion resistant. It is often alloyed with zinc to form brass. Figure 8.4 is the copper-rich end of the copper-zinc phase diagram. Commercial alloys contain up to about 40% zinc. Figure 4.12 shows the increase of strength of brass with zinc content. Because zinc is less expensive than copper, brass is cheaper than pure copper. The architectural use of brass depends on its color.

An alloy of copper with tin is called *bronze*; however, the term bronze has been used for other alloys, such as aluminum bronze, silicon bronze, and manganese bronze. Very few copper-base alloys are hardenable by heat treatment because precipitation from solid solution is not possible. Brass and aluminum bronze do harden rapidly, however. The only important precipitation-hardenable copper alloy

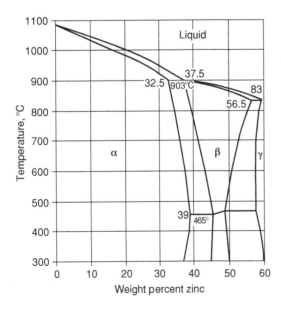

Figure 8.4. Copper-rich end of the Cu-Zn phase diagram. From W. F. Hosford, *Physical Metallurgy*, CRC, 2005.

is beryllium copper containing about 2% Be. Figure 8.5 is the Cu-Be phase diagram. Its use is restricted because of the hazard of berylliosis in its production.

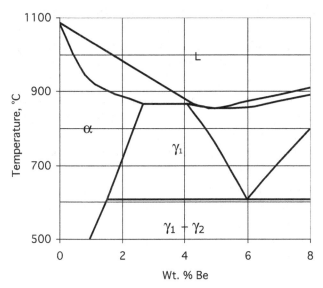

Figure 8.5. Copper-beryllium phase diagram. Data from American Society for Metals, *Metals Handbook*, 8th ed., v. 5, 1973.

Magnesium

The density of magnesium is about 2/3 that of aluminum and 1/4 that of steel. Its low density accounts for most of the use of magnesium-base alloys. Because of their hexagonal close-packed crystal structure, magnesium alloys have limited ductility and are most often used as castings. Magnesium's use in automobiles, kitchen appliances, and garden tools is increasing. Injection molding of semi-solids (*thixomolding*) is finding application for laptop computers, cameras, and cell phones. Magnesium is commonly alloyed with zinc and aluminum. Both have appreciable solubility above 300 °C and decreasing solubility at lower temperatures, which permits precipitation hardening. Other alloying elements include thorium, manganese, zirconium, and rare earths.

Titanium

Titanium has a density ($\rho = 4.51$) between that of aluminum and steel. Its uses are mostly for aerospace, and chemical and sporting goods applications. The aerospace and sporting goods uses are based on its high strength-to-weight ratio. Its use in the chemical industry depends on its corrosion resistance to saline solutions. The biggest disadvantage of titanium for widespread use is its high cost. Titanium undergoes a phase transformation from α (hcp) to β (bcc) as it is heated above 883 °C. This permits alloys to be hardened by heat treatment.

Titanium is produced by reduction of titanium oxide by sodium. The resultant product is a sponge that must be melted to produce useful products. Because titanium readily dissolves both oxygen and nitrogen, it must be processed in a vacuum or an inert atmosphere. This includes melting as well as hot working. Because of this requirement, the cost of finished titanium parts is very high relative to the cost of sponge.

Alloying elements in titanium can be classified by whether they stabilize the α phase or the β phase (e.g., whether they raise or lower the temperature at which $\alpha \rightarrow \beta$). Interstitial oxygen and nitrogen have much higher solubilities in hcp α than bcc β, and are called α stabilizers. Carbon, which also dissolves interstitially, is a weaker α stabilizer. The only other important α stabilizer is aluminum. Tin and zirconium are almost neutral with regard to stabilizing either phase. Aluminum and tin are strong solid solution strengtheners. The bcc elements vanadium, chromium, molybdenum, tantalum iron, and niobium are all β stabilizers.

Some of the important titanium alloys are listed in Table 8.4.

Table 8.4. *Titanium alloys*

Grade	N(max)	C(max)	O(max)	Al	Sn	Zr	Mo	Other
Alpha								
ASTM 1	0.03	0.10	0.18	–	–	–	–	–
Ti code 12	0.03	0.01	0.25	–	–	–	0.3	0.8 Ni
Ti-5Al-2.5Sn	0.05	0.08	0.20	5	2.5			
Ti-8Al-1Mo-1V				8	–	–	1	1V
Ti-6Al-2Nb-1Ta-0.8Mo				6			0.8	2 Nb, 1 Ta
Alpha-Beta								
Ti-6Al-4V				6	–	–	–	4 V
Ti-6Al-2Sn-2Zr-2Mo-2Cr				6	2	2	2	2 Cr, 0.25 Si
Ti-10V-2Fe-3Al				3	–	–	–	10 V, 2 Fe
Beta								
Ti-13V-11Cr-3Al								
Ti-3Al-8V-6Cr-4Mo-4Zr				3	–	4		8 V, 6 Cr
Ti-11.5Mo-6Zr-4.5Sn				–	4.5	6	11.5	

Nickel-base Alloys

Nickel has an fcc crystal structure. Its melting point is 1455 °C and its density is 8.9. Its main uses are for heat-resisting alloys including the super alloys, in magnetic alloys, and in the chemical industry for corrosion resistance,

Relatively pure Ni with or without small amounts of Mn, Al, and Si has a very good corrosion resistance in air and seawater. *Monel* contains 67% Ni-33% Cu, which is approximately the Ni/Cu ratio in Sudbury Ontario deposits. Monel finds widespread use in the chemical industry because of its corrosion resistance. Other copper-nickel alloys include *cupronickel* (25% Ni) and *constantan* (45% Ni). Constantan is used in thermocouples and precision resistors because its resistivity is independent of temperature.

The solubility of chromium in nickel is over 30%. Chromium additions provide oxidation resistance at elevated temperatures. Nickel-chromium alloys are widely used for electrical heating elements. These contain typically 20% Cr and 1.2 to 2% Si with or without smaller amounts of Fe and Al. These are used for furnace windings up to 1200 °C as well as for toasters. *Nichrome*, which contains 20% chromium, is used for resistance heating elements. Nickel-base superalloys containing chromium, cobalt, aluminum, and/or titanium are used for their high temperature strength and oxidation resistance in turbines and jet engines.

Zinc

The largest usage of zinc is for galvanizing steel to inhibit corrosion. The principal use of zinc-base alloys is in the die-casting of small parts.

Notes of Interest

The first precipitation-hardenable alloy is duralumin, which contains 3 or 4% copper, 0.5% to 1% manganese and 0.5% to 1.5% magnesium. Duralumin is a strong, hard, lightweight alloy of aluminum, widely used in aircraft construction. It was discovered accidentally in 1910 by German metallurgist Alfred Wilm during a series of experiments in which it was heated and quenched to see if it would harden like steel during quenching. Hardness measurements were interrupted by a long weekend. Wilm found that the hardnesses measured after the weekend were higher that those immediately after quenching. Its name comes from the town of at Düren in Germany. Its first use was in German-built dirigibles used to bomb England during the first world war.

Titanium was discovered by William Gregor in England in 1791. However, titanium metal was a rarity until 1946 when W. J. Kroll found a process for producing it commercially by reducing titanium tetrachloride with magnesium in what is now known as the Kroll process.

Seven metals are mentioned in the Bible: gold, silver, copper, iron, lead, mercury (quicksilver), and tin. Their chemical symbols (Au, Ag, Cu, Fe, Pb, Hg, and Sn, respectively) come from their Latin names, a testament to their antiquity. Probably the first metals discovered were gold and silver, and perhaps copper. These metals occur in nature in the metallic form, although copper does so rarely. Metallic iron is found only in meteorites.

The earliest use of metals was about 8000 B.C. in Iran and Turkey. Between 4000 and 2000 B.C., metallurgy was limited to copper, silver, gold, and tin. The first smelting occurred around 4000 B.C. Iron smelting began in India about 1200 B.C.

The earliest tin-alloy bronzes date to the late 3000s B.C. in Iran and Iraq. The first bronze probably occurred by accidentally smelting a mixture of copper and tin ores. The development of bronze tools was essential to the growth of civilization.

Problems

1. Aluminum alloy 2017 is used for rivets. These must be kept refrigerated before use. Explain why.
2. Describe a heat treating schedule including times, temperatures, and cooling rates to produce a maximum hardness in aluminum alloy 2024 that contains 4.5% Cu. Neglect the presence of other alloying elements. Production requires that the aging time be no longer than two hours. Figure 8.1 is the Al-Cu phase diagram and the aging characteristics of 2014 aluminum are shown in Figure 8.2.
3. Alpha titanium has an hcp crystal structure with $a = 0.2950$ nm and $c = 0.4683$ nm at 20 °C. The average coefficient of thermal expansion between 20 ° and 900 °C is 8.4×10^{-6}/K. At 900 °C it transforms to β, a bcc crystal structure with $a = 0.332$ nm. Calculate the volume associated with this transformation and indicate which phase is denser.
4. Figure 8.6 shows the microstructure of a piece of Ti-4Al-4V slowly cooled from 1000 °C. Deduce which is the darker phase.
5. A part is presently being made from aluminum alloy 356. If it is made from magnesium alloy AZ91, by how much would the material cost increase or decrease? Assume the same dies are used and that the costs of alloy 356 and AZ91 are $1.12 and $1.65 per pound, respectively.
6. Specify the temperatures, times, and cooling rates necessary to achieve maximum hardness in a part of aluminum alloy 2014. Production schedules limit aging times to two hours. Temperature variations within a furnace are ± 5 °C. Neglect the minor alloying elements and assume that 2014 contains 4.5% copper.

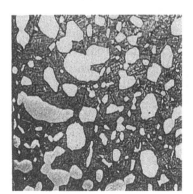

Figure 8.6. Microstructure of Ti-4Al-4V slowly cooled from Fr 1000 °C. From American Society for Metals, *Metals Handbook*, 8th ed., v. 7, 1972.

9 Casting and Welding

Casting

The production of most useful metallic objects involves casting, whether in final form or as ingots that are later shaped as solids by rolling, extruding, or forging. Cast metal components include engine blocks and suspension parts for railcars, trucks, and autos; valves, pumps, faucets, pipes, and fitting equipment for drilling oil wells; surgical equipment and prosthetic devices; and components for household and electronic devices. Injection molding, which is a form of casting, is used to produce many polymer objects.

A number of considerations are important in casting. These include liquid-to-solid shrinkage that requires a reservoir or *riser* of liquid to prevent void formation; thermal shrinkage of the solid that must be accounted for in designing molds; thermal gradients that can cause warping and residual stresses; segregation of components in solution and gas evolution during freezing; and surface appearance. Technological advances in the past few decades have improved the quality and decreased the cost of castings. Computer analysis has allowed prediction and control of the flow of molten metal in the mold, the temperature profiles, and the position of the solid-liquid interface during solidification. The result is the possibility of elimination of internal voids. The use of styrofoam patterns has increased productivity by simplifying mold making.

Macrostructure of Castings

Typically the outside skin of a casting is composed of fine grains of random orientation. As freezing progresses inward, grains that are more favorably oriented for growth crowd out less favorably oriented grains and form *columnar* crystals. Figure 9.1 shows the structure of a cast ingot. Sometimes randomly oriented grains are found near the center.

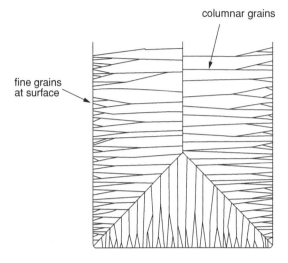

columnar grains

fine grains
at surface

Figure 9.1. Structure of an ingot showing fine grains on the outside and columnar crystals in the interior.

Dendrite Formation

Pure metals freeze with a plane front of solid advancing into the liquid. Alloys, on the other hand, often freeze by *dendritic* growth. Figure 9.2 schematically illustrates the difference between plane front and dendritic growth.

Liquid-to-solid Shrinkage

For most metals, the solid is denser than the liquid and therefore occupies less volume. Table 9.1 shows the shrinkage of several important metals.

The shrinkages for polymers are somewhat larger. For example, polyethylene shrinks about 10% on freezing.

Graphite forms during the freezing of cast iron. The low density of graphite compensates for the shrinkage of the iron matrix. The net lack of shrinkage makes cast iron much easier to cast than most other metals. High silicon contents of cast aluminum alloys have a similar effect.

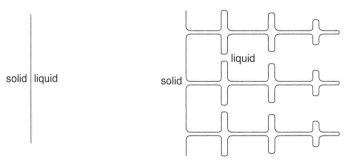

solid | liquid

solid

liquid

Figure 9.2. Plane-front growth (left) and dendritic growth (right).

Table 9.1. *Volume shrinkage on freezing*

Metal	Crystal Structure	% Vol. Change on Freezing	Metal	Crystal Structure	% Vol. Change on Freezing
Mg	hcp	4.1	Zn	hcp	4.2
Sn	bct	2.8	Fe	bcc	3.4
Al	fcc	6.0	Cu	fcc	4.15
Ag	fcc	3.8	Pb	fcc	3.5

Example Problem 9–1:

Pure aluminum shrinks 6% as it solidifies. For an Al-18% Si alloy there is no shrinkage. Estimate the shrinkage for a eutectic Al-Si alloy.

Solution: Figure 8.3 shows that the eutectic is at 12.6% Si. Assuming a linear relation, % shrinkage $= 6(1 - \%\text{Si}/18)$. Substituting % Si $= 12.6$, % shrinkage $= 1.8\%$.

Voids will result in castings unless the freezing is made to occur directionally so that additional liquid can be continuously fed into the unfrozen region. Figure 9.3

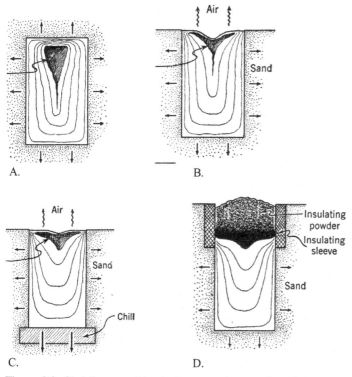

Figure 9.3. Shrinkage cavities in ingot casting can be minimized by promoting directional freezing. A casting completely imbedded in sand (A) has a large shrinkage cavity. Allowing air cooling at the top (B) reduces the shrinkage cavity. Use of chills at the bottom (C) or insulating the top (D) almost completely eliminates the shrinkage cavity. From J. Wulff, H. F. Taylor, and A. J. Schaler, *Metallurgy for Engineers*, Wiley, 1952, p. 409.

shows how insulating the top of a casting and chilling the bottom can prevent shrink-age cavities in ingot casting.

For casting of shapes, *risers* to supply additional liquid and the use of chills are ways of eliminating internal shrinkage cavities. The risers must be large enough so they will not freeze. The riser will later be cut off the casting and remelted. *Chills* are metal or graphite blocks used to absorb heat and hasten freezing at specific locations. Figure 9.4 shows the use of chills and risers to prevent a shrinkage cavity inside a casting.

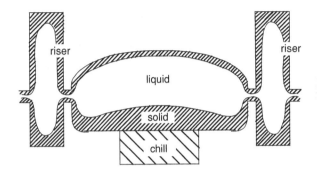

Figure 9.4. Use of risers and chills to pro-duce a solid casting in aluminum.

With dendritic growth, shrinkage porosity may occur interdendritically instead of forming large voids.

Solid-state Shrinkage

Shrinkage as the solid cools after freezing affects the final dimensions of the casting. To allow for this shrinkage, the dimensions of the mold must be made larger than the desired casting by 1 or 2%. Shrinkage must not be constrained, or parts will crack during freezing. This is called *hot tearing*. Figure 9.5 shows how this might result. If one region of a casting cools faster than another, stresses may cause local plastic deformation. This can cause either warping or residual stresses.

Segregation

The first solid to form as an alloy freezes is purer than solid that forms later. This is illustrated in Figure 9.6. Diffusion is not fast enough to eliminate concentration gradients that form on freezing. If there is dendritic growth, the segregation will be

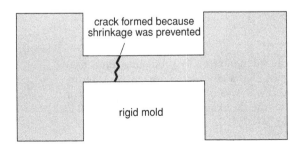

Figure 9.5. Hot tearing caused by restrain-ing shrinkage in the solid state.

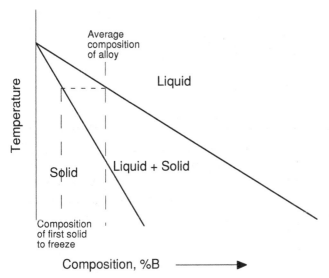

Figure 9.6. Partial phase diagram illustrating why the first solid to form is purer than material that solidifies later.

between the centers of dendrites in the interdendritic regions. Otherwise segregation may be on a much larger scale, with the surface of the casting having a different composition than the interior.

Example Problem 9–2:

Figure 9.7 is the copper-tin phase diagram. Consider the composition of a copper-base alloy containing 5% tin. What is the composition of the first solid to form? Realizing that the initial freezing of relatively pure copper enriches the liquid in tin, estimate the formation of the composition of the last solid to form.

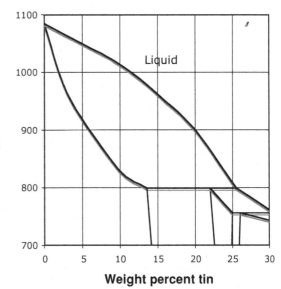

Figure 9.7. Copper-rich end of the copper-tin phase diagram.

Solution: The first solid will contain about 1% Sn. As freezing progresses, the liquid composition will become enriched to nearly 25% Sn. This will undergo a peritectic reaction with the α to form an intermediate phase with about 22% Sn.

Gas Evolution

The solubility of gasses in liquids is much higher than in solids, so dissolved gas will be released as the metal freezes. Released gas may form large voids with plane-front growth. With dendritic growth, the gas is likely to be trapped interdendritically. Aluminum dissolves hydrogen readily. Usually hydrogen must be removed by bubbling insoluble nitrogen through the molten metal before pouring. Aluminum is added to low-carbon steel to heat with disolved oxygen, and thereby prevent violent evolution CO_2.

Casting Processes

Metals are often cast into sand molds. The resulting surface can be no smoother than the sand. In the past, molds were almost always made from new sand, but today the rate of recycling of sand exceeds 90%. Permanent metal molds can be used with metals having lower melting points. Metal molds produce much better surfaces. In other cases ceramic molds may be made around a wax pattern by repeatedly dipping the pattern into a fine ceramic slurry. Metal is poured in after the wax has been melted out. This *lost wax* process finds application in fields as divergent as casting superalloy turbine blades, jewelry, and dental fillings. A lost foam process involves packing sand around a styrofoam pattern, which has the shape of the desired casting. The stryofoam vaporizes as the hot metal is poured in.

Die casting of metals with low melting points (principally aluminum, zinc, and magnesium alloys) and injection molding of polymers are similar processes. Molten material is forced under pressure into a closed water-cooled metal mold. As soon as the metal or polymer is hard enough, the mold opens and the part is ejected. The capital cost of the machinery is high, but the high rate of production makes the process economical where a large number of parts are made. The important properties of aluminum, magnesium, and zinc die-casting alloys are listed in Table 9.2.

Table 9.2. *Properties of three die-casting alloys*

	Aluminum	Magnesium	Zinc
Alloy	Al 360	AZ91	Zamak-3
Composition	9.5 Si, 0.5 Mg	9 Al, 1 Zn	4.3 Al%
T(liquidus)	595	595	307 °C
T(solidus)	555	470	301 °C
Density	2.63	1.81	6.6 Mg/m^3
Heat of fusion	389	373	101 kJ/kg
Specific heat	0.963	1.05	0.419 kJ/kg. °C
Ejection temperature	450	400	250 °C

Thixomolding® is a relatively new process in which an alloy is heated into the two-phase liquid-solid region and injected into a steel mold to form a complex part. It is finding use with magnesium alloys because it is inexpensive, can form very thin sections, and provides large weight savings.

Fusion Welding

Welding is the joining of two pieces of metal or thermoplastic into a single piece. Most welding processes involve melting, though some welding processes are done without. The sources of heat for fusion welding include electrical arcs (AC or DC), flames, lasers or electron beams, and electrical resistance.

In arc welding it is usually necessary to protect the molten metal from the atmosphere. Cellulose coatings on electrodes in DC arc welding decompose to form carbon monoxide and hydrogen, which protect the weld metal. Figure 9.8 illustrates this. Metal-inert-gas welding is a process that involves a continuously fed electrode and uses helium or argon to shield the weld.

Most torch welding is done with an oxygen-acetylene flame, though other gasses including hydrogen may be used. A welding rod supplies the filler material. Figure 9.9 shows torch welding. Whether the flame is reducing or oxidizing depends on the fuel-to-oxygen ratio.

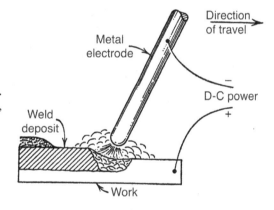

Figure 9.8. Coated electrode. From J. Wulff, H. Taylor, and J. Shaler, *Metallurgy for Engineers*, Wiley and Sons, 1952.

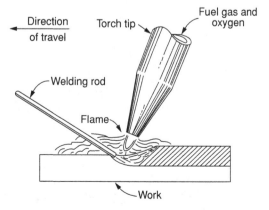

Figure 9.9. Torch welding. From J. Wulff, H. Taylor, and J. Shaler, *Metellurgy for Engineers*, Wiley and Sons, 1952.

The structures of fusion welds are similar to those of castings (Figure 9.10). Freezing results in columnar crystals. There is a heat-affected zone outside of the weld in which grain growth may occur. In an age-hardened alloy, the heat-affected zone will be over-aged. With alloy steels, martensite may form in this region.

In spot welding heat from the resistance of the material to electric current melts a disc-shaped spot (Figure 9.11). Resistive heating is also used in seam and butt welding.

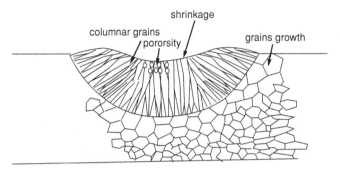

Figure 9.10. Structure of a fusion weld with columnar crystals and grain growth in the heat-affected zone. From W. F. Hosford, *Physical Metallurgy*, CRC, 2005.

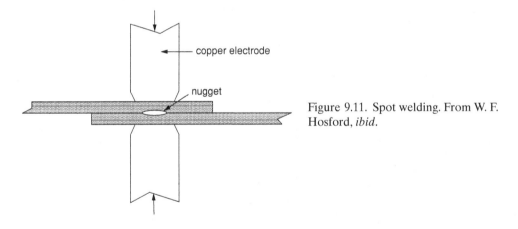

Figure 9.11. Spot welding. From W. F. Hosford, *ibid*.

Solid-state Welding

Direct metal-to-metal contact will cause solid-state welding. Metal-to-metal contact requires plastic deformation to break oxide films. Figure 9.12 shows metal-to-metal contact where oxide films have been broken. Breaking the film may result from indentation that expands the surface, or by friction.

A recent development is the replacement of spot welding by using the friction from rapidly rotating probes to weld sheet metal.

Notes of Interest

The lost wax process dates back many thousands of years. The artists and sculptors of ancient Egypt and Mesopotamia, the Han Dynasty in China, and the Benin

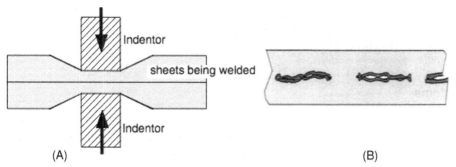

Figure 9.12. Indentation of two sheets (A) causes the metal to deform. This breaks the oxide (B) allowing metal-to-metal contact and welding. From W. F. Hosford, *Physical Metallurgy*, CRC, 2005.

civilization in Africa used the lost wax method of casting to produce their intricately detailed artwork of copper, bronze, and gold.

The Aztec goldsmiths of pre-Columbian Mexico used the lost wax process to create much of their elaborate jewelry. Unfortunately, most of their work was plundered by the conquistadors and melted down into gold bars to enrich the Spanish treasury. The quality of the few surviving pieces demonstrates a mastery of the process that must have taken many years of trial and error to develop. Brass smiths in Benin, Nigeria, still produce lost wax castings using an ancient method. They begin with a core of clay kneaded into a mass. The core of clay is shaped into the approximate size and shape of the article to be made and then allowed to dry for several days. The core is covered with beeswax and formed into the desired shape. It is then covered with a thick coating of clay, the first layer of clay being applied as a very fine slip. The cores may be self-supporting or held in place with small pins. A thin roll of wax is added to form the sprue in through which the molten metal is poured. Subsequent layers of clay are added, gradually creating a mold. This mold is allowed to air dry thoroughly and the wax melted out.

Flywheels have been made since the mid 1800s with curved spokes to avoid hot tearing (Figure 9.13). Curved spokes can accommodate shrinkage by slight unbending with much less stress that straight spokes.

Blacksmiths frequently welded two pieces of wrought iron without melting by pounding them together while they were red hot. To clean the mating faces, silica,

Figure 9.13. Hot tearing is prevented with curved spokes, which can accommodate shrinkage without much stress.

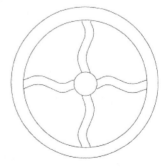

sand, or borax was sprinkled on the surfaces. This formed a low-melting eutectic with the iron oxide, which was squeezed out when pressure was applied, allowing metal-to-metal contact.

Problems

1. Pure iron has a density of 7.86 Mg/m^3; the density of graphite is 2.25 Mg/m^3. Calculate the % C in a ferritic gray cast iron with a density of 7.09 Mg/m^3. Assume the structure consists of pure iron and graphite.

2. The thermal expansion coefficient of aluminum alloys is $22 \times 10^{-6}/°C$. How much larger than the final casting must the mold be?

3. Why does tin shrink less on solidifying than the other metals in Table 9.1?

4. If a precipitation-hardened alloy is welded, there is likely to be a region just outside of the weld that is weaker than the rest of the metal. Explain.

5. For parts of the same volume, and with the same cooling, which of the three die-casting alloys shown in Table 9.2 would have the shortest cycle time in die casting? The latent heats of freezing are 397 J/g for aluminum, 101.2 J/g for zinc, and 193 J/g for magnesium.

6. Which of the three cooling curves in Figure 9.14 is appropriate for an aluminum alloy containing 5% Si?

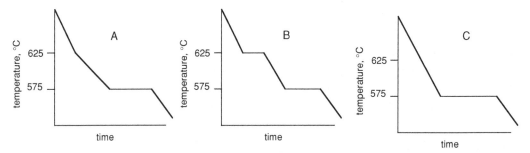

Figure 9.14. Possible cooling curves.

10 Solid Shaping

Bulk Forming

Processes for forming solid metals can be classified as bulk forming in which the forces are mostly compressive, and sheet forming in which the metal is stretched in tension. Bulk forming processes may be further classified as either hot working or cold working, depending on whether the work material leaves in a recrystallized state.

Figure 10.1 illustrates several important bulk forming processes. Compression is clearly the dominant force in rolling, extrusion, and forging. For rolling, extrusion, and forging, the reduction is usually limited by the capacity of the machinery to deliver forces. Although drawing involves pulling a rod or wire through a die, the pressure the die exerts on the rod or wire is the dominant force. In rod and wire drawing, the maximum reduction per pass is limited by the possibility of tensile failure of the drawn wire. Typically a maximum strain per pass is about $\varepsilon = \ln(A_1/A_2) = 0.65$, which corresponds to a diameter reduction of 38%. Obviously many passes are required to make fine wire. Many rolling passes are used to make sheets and shapes. Forging uses repeated blows to achieve a final shape. In contrast, rods, tubes, and other shapes are extruded in a single operation to their final shape because high reductions are possible.

The work per volume, w_a, to change the shape is given by

$$w_a = (1/\eta) \int \sigma \, d\varepsilon, \qquad 10.1$$

where η is a mechanical efficiency, σ is the yield strength, $\varepsilon = \ln(A_o/A_f)$, and A_o and A_f are the original and final cross-sectional areas.

Example Problem 10–1:

Calculate the force required to extrude a 20 cm diameter billet of aluminum to a diameter of 2 cm if the yield strength at the extrusion temperature is 10 MPa and the efficiency is 75%. Assume that no work hardening occurs.

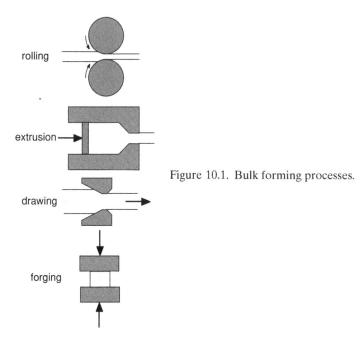

rolling

extrusion

drawing

forging

Figure 10.1. Bulk forming processes.

Solution: With no work hardening $\int \sigma \, d\varepsilon = \sigma \, d\varepsilon \, 10 \times 10^6 \ln(20/2)^2 = 46 \, \text{MJ/m}^3$, so $w_a = 46 \, \text{MJ/m}^3/0.75 = 61 \, \text{MJ/m}^3$. The total work in extrusion is $F_e L$ where F_e is the extrusion force and L is the distance through which the extrusion force works. The work per volume in extruding equals $F_e L/(AL) = 61 \, \text{MJ/m}^3$. $F_e = A(61 \, \text{MJ/m}^3) = \pi(0.10\text{m})^2 = 1.9 \, \text{MN}$.

If the geometry of the deformation zone is such that the ratio, Δ, of material thickness to length of contact with tools is much greater than unity, the deformation will be inhomogeneous. The result will be tensile residual stresses on the surface and the likelihood of interior cracking. On the other hand, if Δ is much less than unity, friction between tools and work piece becomes high.

Hot Working vs. Cold Working

Many bulk forming processes are done hot. Flow stresses decrease at high temperatures so lower tool forces are required. Consequently, equipment size and power requirements are decreased. *Hot working* is often defined as working above the recrystallization temperature so that the work metal recrystallizes as it deforms. However, this is an oversimplified view. The strain rates of many metalworking processes are so high that there is not time for recrystallization to occur during deformation. Rather, recrystallization may occur in the time period between repeated operations, as in forging and multiple-stand rolling, or while the material is cooling to room temperature after the deformation is complete. High temperatures lower the flow stress, whether recrystallization occurs during the deformation or not. Hot-rolled products are in an annealed state.

The elevated temperatures during hot working have several undesirable effects. Among them are:

1. Lubrication is more difficult. Although viscous glasses are often used in hot extrusions, most hot working is done without lubrication.
2. The work metal tends to oxidize. Scaling of steel and copper alloys causes loss of metal and roughened surfaces. While processing under inert atmosphere is possible, it is prohibitively expensive and is avoided except in the case of very reactive metals, such as titanium.
3. Tool life is shortened because of heating, the presence of abrasive scales, and the lack of lubrication. Sometimes damage is minimized by the use of scale breakers and cooling the rolls by water spray.
4. Poor surface finish and loss of precise gauge control result from the lack of adequate lubrication, oxide scales, and roughened tools.
5. The lack of work hardening is undesirable where the strength level of a cold-worked product is needed.

Because of these limitations, it is common to hot-roll steel to about 0.08 in. (2 mm) thickness to take advantage of the decreased flow stress at high temperature. The hot-rolled product is then pickled in acid to remove scale, and further rolling is done cold to ensure good surface finish and optimum mechanical properties. The cold-rolled steel sheet is almost always sold in an annealed state. Annealing is done in a controlled atmosphere after cold rolling. The principal advantage of cold-rolled sheet steel over hot-rolled sheet is a much better surface finish.

Sheet Forming

Cups can be formed by forcing discs cut from sheets through dies as illustrated in Figure 10.2. The maximum diameter reduction is limited by failure of the cup walls. This usually limits the height-to-diameter ratio to about 0.75.

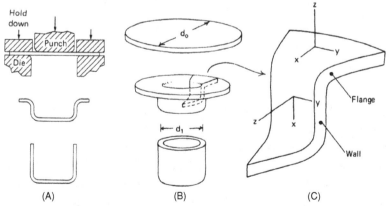

Figure 10.2. Schematic illustration of forming a cup from sheet metal. A descending punch forces the sheet through a circular hole in the die (A). A hold-down plate prevents buckling. The major deformation is the circumferential contraction of the flange so that it can pass over the die lip (B). By focusing on a pie-shaped segment, it can be seen that the flange is under hoop compression and the wall under tension. From W. F. Hosford and R. M. Caddell, *Metal Forming: Mechanics and Metallurgy*, 3rd ed., Cambridge, 2007.

Example Problem 10–2:

In cup drawing, the largest possible drawing ratio, d_0/d_1, is about 2.0. Assuming there is no thinning or thickening of the sheet as it is drawn, what is the maximum height-to-diameter ratio that can be achieved?

Solution: With no thinning or thickening, the surface area is constant, so $\pi d_0^2/4 = \pi d_1^2/4 + \pi d_1 h$. Substituting $d_0 = 2d_1$, $4d_1^2 = d_1^2 + 4d_1 h$, $d_1/h = 0.75$.

Deeper cups can be formed by *redrawing* and/or *ironing*. Figure 10.3 illustrates redrawing a cup to a smaller diameter and increased height. Redrawing causes little change of wall thickness. The height of drawn or redrawn cups can be further increased by ironing, which thins the wall. Figure 10.4 shows how an ironing ring set into the die wall can force the wall to thin and elongate.

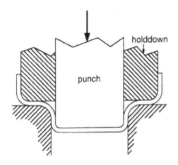

Figure 10.3. Redrawing a cup to a smaller diameter.

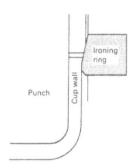

Figure 10.4. Ironing. An ironing ring thins the cup wall and increases the cup height. From W. F. Hosford and R. M. Caddell, *Metal Forming: Mechanics and Metallurgy*, 3rd ed., Cambridge, 2007.

Products made by drawing include beverage cans, dry cells, cartridge cases, propane cylinders, cooking pots, and oil filters.

Many sheet forming processes stretch the sheet. In the production of complex shapes such as auto bodies, sheets are clamped at their edges and stretched to a final shape by a shaped punch as illustrated in Figure 10.5. Draw beads control how much material from the flange is drawn into the die. Wrinkling of the finished part results if too much material is allowed to flow into the die. It should be emphasized that in stamping processes the sheet is not squeezed between the die and punch. Rather tensile forces in the sheet cause it to deform by stretching. If too much stretching occurs, the sheet may suffer a tensile failure in the die.

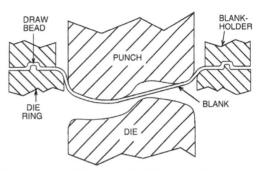

Figure 10.5. Stamping of a sheet metal part. Note that the restraint from the draw beads in the blank holder promotes more stretching of the sheet. From W. F. Hosford, *Mechanical Behavior of Materials*, Cambridge, 2005.

The main problems encountered in stamping are wrinkling, tensile failure, and springback (partial elastic recovery so the final shape is not retained). Surface appearance is also of concern. Increased blank holder pressure causes greater tensile stretching of the sheet, which tends to decrease wrinkling and springback, but increases the tendency for tensile failure.

Note of Interest

Aluminum foil is produced by rolling two sheets of aluminum simultaneously. When the sheets are separated, the sides that were in contact with each other have a matte finish, the roughness of which reflects the grain size. The other sides are shiny because of sliding on the polished steel rolls.

Problems

1. A typical beverage can is 2.6 in. in diameter and 4.8 in. high. The thickness of the bottom is 0.010 in. and the wall thickness is 0.0035 in.
   a. Assuming that the blank thickness was 1.010 in., calculate the diameter of the original blank.
   b. Calculate the strain at the top of the can.
2. A sheet of steel 2.5 mm thick is rolled to a thickness of 1.0 mm. Assuming that the required stress is 175 MPa and that the deformation is adiabatic (all of the energy required to deform the steel is converted to heat), calculate the temperature rise. Data for the steel: density $= 7.88$ Mg/m^3, heat capacity $= 449$ J/kg-K.
3. Discuss whether making a series of light rolling reductions instead of one large reduction will result in greater homogeneity and lower residual stresses.
4. If the maximum strain per pass in a wire-drawing operation is $\varepsilon = \ln(A_0/A) = 0.72$, how many drawing passes would be required to draw a wire from 5 mm diameter to 0.2 mm diameter?

11 Polymers

Polymer is the technical term for what are commonly called plastics. Polymers consist of very large molecules. In fact, the word polymer comes from *poly* (for many) and *mer* (for parts). There are two main groups of polymers: thermoplastics, which are composed of linear molecules and will soften and melt when heated, and thermosetting polymers, whose molecules form three-dimensional framework structures. Thermosetting polymers will not melt or even soften appreciably when heated. The basic organic chemistry of bonding in polymers is treated in Appendix 7.

Thermoplastics

Thermoplastics consist of long chain molecules. The simplest of these have carbon-carbon backbones. The structure of polyethylene is illustrated schematically in Figure 11.1. Other linear polymers with carbon-carbon backbones are listed in Table 11.1.

Other thermoplastics have more complex backbones. Among these are nylon, polyester (PET), polycarbonate, cellulose, and poly(paraphenylene terephthalamide) or PPTA (also known as Kevlar®). Figure 11.2 illustrates the molecular structure of these.

Elastomers are rubber-like polymers. They have carbon backbones that include carbon-carbon double bonds. Figure 11.3 shows the repeating unit. The radicals, R, are listed in Table 11.2.

Silicones are polymers with -Si-O-Si-O- backbones.

Degree of Polymerization

The degree of polymerization is the average number of mer units per molecule. It is the molecular weight of the polymer divided by the molecular weight of the mer. There are two ways of describing the molecular weight: the number-average molecular weight, and the weight-average molecular weight. They are usually not very different but the weight average is always somewhat larger. The difference is treated in Appendix 8.

Figure 11.1. Molecular structure of polyethylene. The black
dots represent hydrogen atoms.

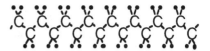

Table 11.1. *Polymers of general form* $\begin{array}{c} \\ {-}\overset{\overset{1}{|}}{\underset{\underset{2}{|}}{C}}{-}\overset{\overset{3}{|}}{\underset{\underset{4}{|}}{C}}{-} \end{array}$

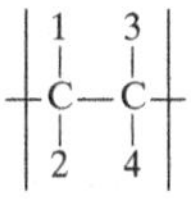

Polymer name	1	2	3	4	Abbreviation
Polyethylene	H	H	H	H	PE
Polyvinyls (general)	H	H	H	R	
Polyvinyl chloride	H	H	H	Cl	PVC
Polyvinyl alcohol	H	H	H	OH	PVAl
Polypropylene	H	H	H	CH_3	PP
Polyvinyl acetate	H	H	H	$O{-}\overset{\overset{O}{\|}}{C}{-}CH_3$	PVAc
Polystyrene	H	H	H	 $\langle\!\langle\ \rangle\!\rangle$	PS
Poly acrylonitrile	H	H	H	N	PAN
Polyvinylidenes (general)	H	H	R_1	R_2	
Polyvinylidene chloride	H	H	Cl	Cl	PVDC
Polymethylemethacrylate	H	H	CH_3	$O{-}\overset{\overset{O}{\|}}{C}{-}C{-}CH_3$	PMMA
Polyisobutylene	H	H	CH_3	CH_3	PIB
Polytetrafluoroethylene	F	F	F	F	PTFE

Table 11.2. *Several rubbers*

R	rubber
H	polybutadiene (buta-rubber)
Cl	polychloroprene
CH_3	polyisoprene (natural rubber)

Example Problem 11–1:

Calculate the degree of polymerization of polyethylene having a molecular weight
of 10,000.

Solution: The degree of polymerization (DP) is the number of mers per molecule and
equals the molecular weight divided by the mer weight. For polyethylene, the mer
contains 2 carbon atoms and 4 hydrogen atoms, so the mer weight is $2 \times 12 + 4 \times 1 = 28$. $DP = 10,000/28 = 357$.

Molecular Weight

Both strength and melting temperature increase with molecular weight. Figure 11.4
shows this effect for the paraffin series, C_xH_{2x+2}.

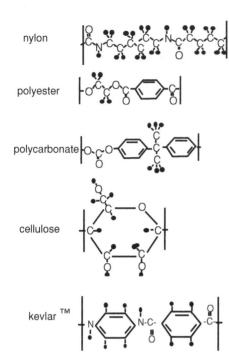

nylon

polyester

polycarbonate

cellulose

kevlar ™

Figure 11.2. Structure of several linear thermoplastics without C-C backbones. The black dots represent hydrogen atoms.

Figure 11.3. Basic structure of rubbers. The filled circles are hydrogen atoms, and R stands for a radical.

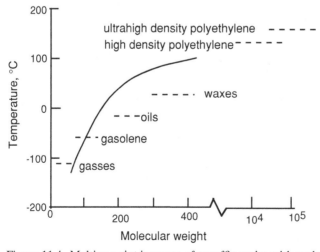

Figure 11.4. Melting point increase of paraffin series with molecular weight.

Branching

Linear polymers may form branches, as illustrated in Figure 11.5.

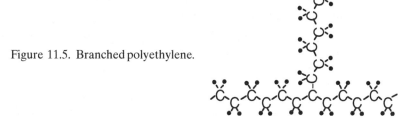

Figure 11.5. Branched polyethylene.

Cross-linking

Cross-links may be formed between molecular chains. Rubber is vulcanized with sulfur to form cross-links between polyisoprene molecules as shown in Figure 11.6. The hardness of the rubber increases with the number of cross-links.

Figure 11.6. Cross-linking of rubber molecules by sulfur.

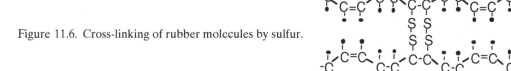

Stereoisomerism

There may be more than one distinguishable arrangement of side groups around the carbon-carbon chains in molecules of the same chemical formula. Rotation about the carbon-carbon bond cannot change these distinguishable arrangements. One example is polypropylene. If all of the CH_3 groups are on the same side of the chain, it is said to be *isotactic*; if they are on opposite sides, it is said to be *syndiotactic*. See Figure 11.7. If the position of the CH_3 groups is random, the polymer is *atactic*.

isotactic

atactic

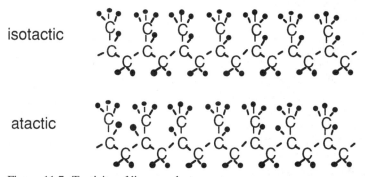

Figure 11.7. Tacticity of linear polymers.

Copolymers

Many commercial plastics are copolymers. Two or more monomers are polymerized together. ABS is a copolymer of acrylonitrile, butadiene, and styrene that incorporates some of the rubber-like properties of buta-rubber. What is commercially called vinyl is a copolymer of polyvinyl acetate and polyvinyl chloride. The properties of vinyl depend on the proportions of vinyl acetate and vinyl chloride as indicated in Table 11.3.

Table 11.3. *Polyvinyl chloride-vinyl acetate copolymers*

Polymer	W/O vinyl acetate	Typical MW	Applications
Pure vinyl chloride	0		siding, pipes
PVC-PVA copolymer	5	20,000	insulating wiring, must be plasticized
	10 to 12	16,000 to 23,000	fibers for fabric
	13 to 15	10,000	injection molded plastics
	13 to 15	9,000	lacquer for food cans, surface coating
Pure polyvinyl acetate	0	5,00 to 15,000	adhesives

The monomer units may be distributed randomly or occur in blocks as illustrated in Figure 11.8.

AAABABABBABABBABAAAABBABAABBABABBABBABABB

AAAAAAAAAAAAABBBBBBBBBBBBBBBBBBBBBAAAAAAAAAA

Figure 11.8. Random copolymer (top) and a block copolymer (bottom).

With ABS the butadiene rubber is in the form of small spherical particles as shown in Figure 11.9.

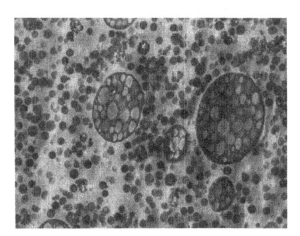

Figure 11.9. Microstructure of ABS. Note the dark spheres of butadiene rubber. From *Engineering Materials Handbook, v. 2, Engineering Plastics*, ASM International, 1988.

Molecular Configuration

Linear polymers may be either amorphous or crystalline, or may contain both amorphous and crystalline regions. One may think of the molecules in an amorphous polymer as being like cooked spaghetti. In crystalline regions the molecules are parallel and fit closely together. Figure 11.10 illustrates these two possibilities.

Figure 11.10. In amorphous regions (left) the molecules are randomly oriented while in crystalline regions the molecules are aligned.

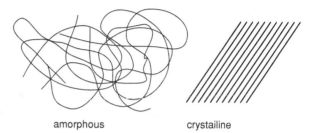

amorphous   crystailline

The crystalline regions have a higher density and are stiffer than the amorphous regions. The factors that tend to prevent crystallinity are those that decrease the order of the polymer chains such as side branching, cross-linking, large side groups, irregular spacing of side groups (being atactic), and copolymerization. As a linear polymer is stretched, the molecules tend to become aligned with the direction of stretching. This increases the crystallinity and increases the stiffness.

Some polymers are highly crystalline, but others have little or no crystallinity. Branching, large side groups, and copolymerization all tend to suppress crystallinity. There is no branching in high-density polyethylene, but low-density polyethylene is highly branched. The crystallinity of several polymers is listed in Table 11.4.

Table 11.4. *Crystallinity of several polymers*

Polymer	% Crystalline
High-density polyethylene	90
Low-density polyethylene	40
Isotactic polypropylene	90
Atactic polypropylene	0
Polystyrene	0
Trans and cis polybutadiene	80
Randon cis-trans polybutadiene	0

Thermosetting Polymers

The molecules of thermosetting polymers form three-dimensional networks. Thermosets are formed by two sets of monomers, one of which must be able to react in more than two places (i.e., be *polyfunctional*). Figure 11.11 is a schematic sketch of such a network. The open circles represent monomers that are trifunctional. Some have reacted with three smaller monomers (small filled circles) while others have reacted with only two.

Figure 11.11. Schematic illustration of a three-dimensional network formed by two types of monomers. The open circles represent a large monomer and the black circles another smaller monomer.

Among common thermosetting polymers are phenol formaldehyde, illustrated in Figure 11.12, and urea formaldehyde, illustrated in Figure 11.13. Polyesters and epoxies are also thermosetting.

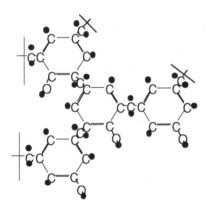

Figure 11.12. Phenol formaldehyde.

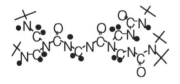

Figure 11.13. Urea formaldehyde.

Addition Polymerization

There are two types of polymerization reactions. One is simple addition polymerization in which similar monomers containing carbon-carbon double bonds react with each other. This must be initiated by breaking the double bond. Ultraviolet light, high temperature, or an initiator (small molecule with a free electron) may cause the double bond to break, forming a free radical that can react with another monomer and causing its double bond to break, forming another free radical. This leads to a chain reaction. The most common way of initiation by addition of a chemical initiator is illustrated in Figure 11.14.

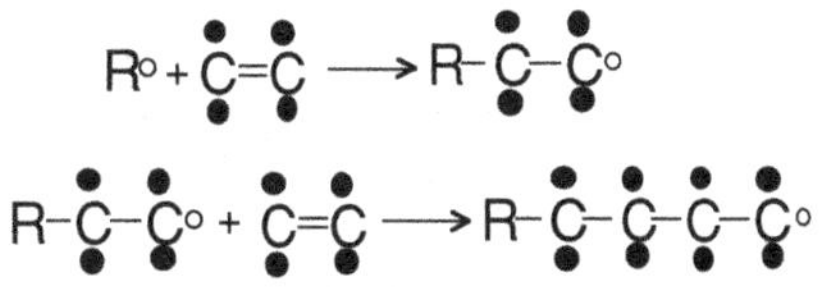

Figure 11.14. Initiation of additional polymerization of polyethylene from ethylene by reaction of an initiator that has a free electron, which causes the double bond to break and create another free electron. Filled circles represent hydrogen atoms and the open circles represent a free electron.

$$R\circ + \overset{\cdots}{\underset{\cdots}{C}}=\overset{\cdots}{\underset{\cdots}{C}} \longrightarrow R-\overset{\cdots}{\underset{\cdots}{C}}-\overset{\cdots}{\underset{\cdots}{C}}\circ$$

$$R-\overset{\cdots}{\underset{\cdots}{C}}-\overset{\cdots}{\underset{\cdots}{C}}\circ + \overset{\cdots}{\underset{\cdots}{C}}=\overset{\cdots}{\underset{\cdots}{C}} \longrightarrow R-\overset{\cdots}{\underset{\cdots}{C}}-\overset{\cdots}{\underset{\cdots}{C}}-\overset{\cdots}{\underset{\cdots}{C}}-\overset{\cdots}{\underset{\cdots}{C}}\circ$$

The molecular weight is controlled by the amount of initiator added. A high molecular weight results only if a little initiator is used.

Condensation Polymerization

Certain reactions between two different monomers result in a small by-product molecule. Often, though not always, the small molecule is H_2O. Figure 11.15 is a schematic illustration of the condensation polymerization of nylon.

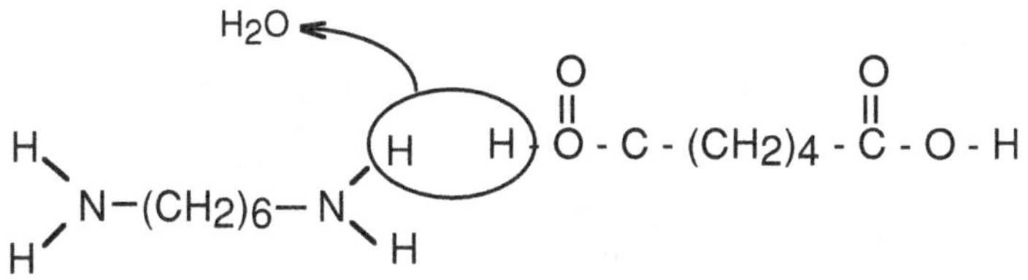

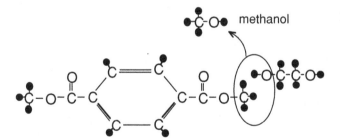

Figure 11.15. Reaction of hexamethyldiamine (left) with adipic acid results in the formation of nylon with water as a by-product. There is room for reaction of only one of the hydrogen atoms at each end of the hexamethyldiamine, so nylon is a linear polymer.

Figure 11.16 shows that methanol is the by-product in the polymerization of polyethylene terephthalate (PET).

Figure 11.16. The by-product of the condensation reaction to form PET is methanol.

Many of the thermosetting polymers are formed by condensation polymerization. The reaction to form phenol formaldehyde and urea formaldehyde releases water.

Example Problem 11–2:

Consider the reaction to form polyester from ethylene glycol and adipic acid (Figure 11.17). What is the by-product of the reaction, and what is the weight of the by-product per weight of polymer formed?

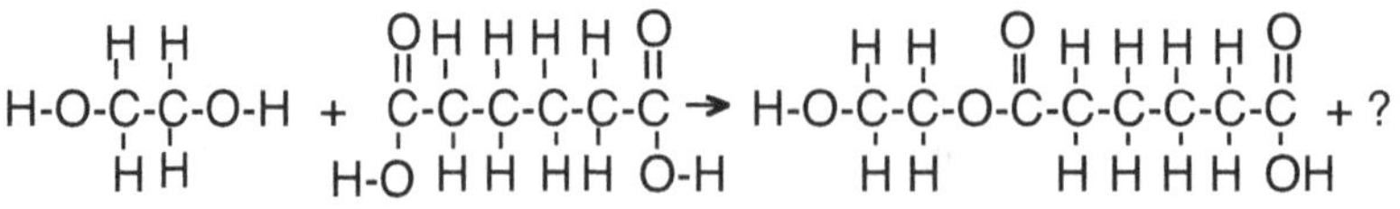

$$\text{H-O-C-C-O-H} + \text{C-C-C-C-C-C} \rightarrow \text{H-O-C-C-O-C-C-C-C-C} + ?$$

Figure 11.17. Reaction of ethylene glycol and adipic acid.

Solution: The by-product is H_2O. The weight of the water is $16 + 2 = 18$ and the weight of the polyester is $5 \times 16 + 8 \times 12 + 14 \times 1 = 190$. The weight of water per polymer is $18/190 = 9.5\%$.

Common Polymers

Table 11.5 lists a number of commercial polymers with their common names, the chemical names, and uses.

Table 11.5. *Commercial polymers*

Common or trade name	Chemical name	Characteristics	Applications
ABS	acrylon-nitrile-butadiene	flexible, shape memory	canoes
Lucite®, Plexiglas®	polymethyl methacrylate	transparent	windows
Nylon	polyvinylidene hexanamine	moldable, low friction	gears, fabric
Saran®	polyvinylidene chloride	transparent thin films	packaging
Dacron®	polyethylene terephthalate	fiber	fabric
Mylar®	polyethylene terephthalate	thin films	electronics
PETE, PET	polyethylene terephthalate	transparent, water resistant	bottles
Bakelite	phenol formaldehyde	moldable	heat resistant insulators
Teflon®	polytetrafluoroethylene	low friction	cooking utensils
Lexan®	polycarbonate	tough	football helmets
Rayon	regenerated cellulose	fiber	fabric
Cellophane	cellulose film	thin film	packaging
Celluloid	nitrocellulose, cellulose acetate	transparent	largely discontinued
Orlon®	acrylic fiber	fiber	fabric
Kevlar®	poly-paraphenylene terephthalamide	extremely strong fiber	reinforcing polymers, bulletproof vests

Glass Transition

The bonds between neighboring molecular chains are weak van der Waals bonds; therefore, most polymers do not readily crystallize. Instead on cooling they form an amorphous glass at a glass transition temperature. The relative positions of molecules are frozen so the amorphous glass is not a liquid. The change of specific volume on cooling is schematically illustrated in Figure 11.18.

The glass transition temperatures of several polymers are given in Table 11.6.

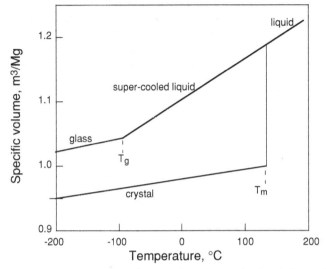

Figure 11.18. Change of the specific volume of polyethylene with temperature. If it does not crystallize at the melting temperature, polyethylene will remain a supercooled liquid until it reaches its glass transition temperature. From W. F. Hosford, *Materials Science: An Intermediate Text*, Cambridge, 2006.

Table 11.6. *Glass transition and melting temperatures of several polymers*

Polymer	$T_g(^\circ C)$	$T_m(^\circ C)$
Polyethylene (high density)	−120	140
Polybutadiene	−70 (±)	*
Polypropylene	−15	175
Nylon 6/6	50	265
Polyvinyl chloride	85	210*
Polystyrene	100	240*

* Difficult to crystallize

At temperatures below their glass transition, polymers tend to have a high Young's modulus and be brittle. Figure 11.19 shows the sharp drop of Young's modulus for polyvinyl chloride (PVC) near its glass transition temperature. Clearly the glass transition temperature is near 80 °C and depends slightly on the time of loading.

The reason that the modulus is time dependent is illustrated by Figure 11.20. For the same stress, the strain increases with the time of loading so the modulus drops.

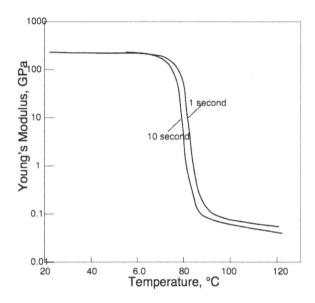

Figure 11.19. The elastic modulus of PVC is about three orders higher in magnitude below the glass transition temperature than above it. It depends only slightly on the rate of loading.

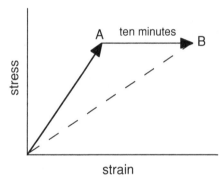

Figure 11.20. At constant stress, the strain increases so the modulus, σ/ε, decreases with time of loading. From W. F. Hosford. *Mechanical Behavior of Materials*, Cambridge, 2005.

Additives

Because the C-C bonds of polymers can be broken by ultraviolet radiation, carbon black, titanium dioxide, and other pigments are often added to block sunlight. Dyes do not perform this function. Fillers such as wood flour and silica are often used to decrease the price. Hard particles such as silica increase the wear resistance. Small molecules called *plasticizers* are added to polymers to decrease the glass transition temperature and elastic modulus as shown in Figure 11.21. In the

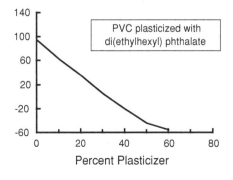

Figure 11.21. Lowering of the glass transition temperature by a plasticizer. From W. F. Hosford, *Materials Science: An Intermediate Text*, Cambridge, 2006.

past, halogen-containing compounds were used as flame retardants, but because of toxicity and environmental concerns these compounds are being replaced by phosphorous-, nitrogen-, and silicon-containing organic compounds, inorganic compounds, and nanomaterials.

Degradation

There is a spectrum of interactions between plastics and solvents. There is no interaction of polyethylene or PET with water. Other plastics will absorb a solvent and swell. Examples include nylon in water and PVC in ketones. Some solvents may dissolve certain polymers. Polyvinyl alcohol dissolves in water. PVC pipes are joined together with solvents that evaporate to form welds. Gasoline will cause hydrocarbon rubbers to swell. Alcohol attacks polyurethanes.

The double bonds of vulcanized rubber are susceptible to ozone attack. The *scission* reaction is illustrated in Figure 11.22.

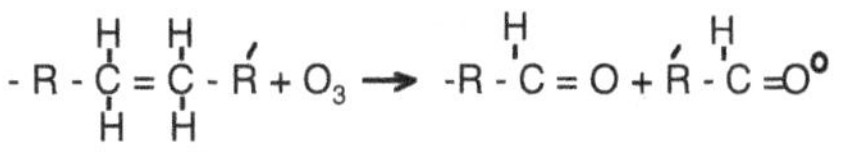

Figure 11.22. Schematic scission reaction breaking a molecule at a double bond.

$$-R-\underset{\underset{H}{|}}{\overset{\overset{H}{|}}{C}}=\underset{\underset{H}{|}}{\overset{\overset{H}{|}}{C}}-R + O_3 \rightarrow -R-\overset{H}{C}=O + R-\overset{H}{C}=O^{\circ}$$

A combination of stress and environments containing some chemicals may lead to stress cracking.

Properties and Uses of Polymers

Polyethylene Terephthalate – PET

Often called polyester.

Its density is 1.35–1.38 g/cm^3; MP $= 260\,°$C, $T_g = 75\,°$C.

It is tough, provides a barrier to water and gasses, and finds use in beverage bottles, food containers, yarns, polyfill, and Mylar® film that is used for its dielectric properties.

High-density Polyethylene – HDPE

Its density is 0.94–0.96 g/cm^3; MP $= 130\,°$C, $T_g = -90\,°$C.

It is highly crystalline giving it a high modulus of elasticity and making it translucent. It is tough, strong, and easy to process, and is resistant to water and chemical attack. It finds use as bottles for milk, bleach, and motor oil.

Low-density Polyethylene – LDPE

Its density is 0.92–0.94 g/cm^3; MP $= 170\,°$C, $T_g = -110\,°$C.

This contains many branches that inhibit crystallization. It is tough, flexible, and transparent, and finds wide usage in wire insulation, squeezable bottles, furniture, and carpets.

Polyvinyl Chloride – PVC

Often called vinyl.

Its density is 1.32–1.42 g/cm^3 (vinyl); MP $= 212\,^{\circ}$C, T$_g = 87\,^{\circ}$C.

It is hard and brittle in unplasticized forms. It is used for building material such as siding, window frames, and piping systems. As plasticized and copolymerized with polyvinyl acetate, it is used for clothing, upholstery, flexible tubing, and non-food containers. Its flexibility depends on how much polyvinyl acetate has been copolymerized with it. It has stable electrical properties, is easy to blend and process, and is suitable for injection molding. PVC is flame resistant but the chlorine produces HCl when PVC is burned.

Polystyrene – PS

Contains a benzene ring.

Its density is 1.05 g/cm^3; MP $= 240\,^{\circ}$C, T$_g = 95\,^{\circ}$C.

It is brittle and hard. It is molded into rigid objects, but the greatest use is as foam for insulation in construction, drinking cups, and packaging (peanuts and rigid shapes).

Polypropylene – PP

Its density varies from 0.85 g/cm^3 for amorphous PP to 0.95 g/cm^3 for crystalline PP; MP $= 165\,^{\circ}$C, T$_g = -10\,^{\circ}$C.

Its resistance to chemical attack makes it suitable for tanks, chemicals, sinks, ducts, barrels, tanks, pump components, and prosthetic devices. As a fiber it is used in "polypro" clothing.

Polymethylmethacrylate – PMMA

Often known by the trade names Lucite®, Plexiglas®, and Perspex®.

Its density is 1.19 g/cm^3; MP $= 130$–$140\,^{\circ}$C, T$_g = 105\,^{\circ}$C.

It is highly transparent. It has a higher impact strength than glass, but is softer and is used wherever toughness transparency is required. Applications include shields for rinks, aircraft windows, colored glasses, and eye lenses.

Nylon 6/6

There are a number of nylon compounds, which vary in the length of the monomers. The most common is Nylon 6/6, which is made from two monomers, hexamethyl diamine and adipic acid, each of which contain six carbon atoms.

Its density is 1.13–1.15 g/cm^3; MP $= 265\,^{\circ}$C, T$_g = 57\,^{\circ}$C.

It is used as a fiber in clothing and rope, as well as molded into gears and other objects that require strength and low friction.

Polycarbonate

Trade name Lexan®.

Its density is 1.20–1.22 g/cm^3; MP $= 267\,^{\circ}$C, T$_g = 150\,^{\circ}$C.

It is very tough and transparent. It is used in applications requiring toughness and heat resistance such as bulletproof glass, injected-molded and extruded shapes, film, signs, and football helmets.

Polyvinylidene Chloride

Its density is 1.7 g/cm^3; MP $= 169\,°$C, $T_g = -17\,°$C.

Its principle use is as Saran wrap.

Aramids – Kevlar®

There are two common grades of Kevlar: Kevlar 29 and Kevlar 49. Its density is 1.44 g/cm^3; MP $= 640\,°$C, $T_g = 375\,°$C.

Kevlar is an aromatic polyamide, typically used for reinforcements in tires and other rubber mechanical goods. Kevlar 29 is used in industrial applications such as cables, replacement for asbestos, brake linings, and body armor. Kevlar 49 is considered to have the greatest tensile strength of all the aramids, and is used in applications such as plastic reinforcement for boat hulls, airplanes, bicycles, bulletproof vests, trampolines, and tennis rackets.

Polytetrafluoroethylene – Teflon®

Its density is 2.2 g/cm^3; MP $= 327\,°$C, $T_g = -110\,°$C. Teflon decomposes at $350\,°$C, so it cannot be molded. Because of its extremely low coefficient of friction it is used for coating armor-piercing bullets, bushings, bearings, and frying pans. Gor-tex®, a breathable polytetrafluoroethylene fabric with micropores, is used for raingear.

Phenol Formaldehyde – Bakelite

Although its use has declined because of cost, it is still used for small parts such as disc cylinders, electric plugs, and switches.

Urea Formaldehyde

It is commonly used for electrical appliance casing such as desk lamps. It has largely replaced bakelite.

Polyurethanes

Polyurethanes are used as both flexible and rigid foams. Flexible foam, often referred to as foam rubber, is used for automobile seats, chairs, mattresses, and rigid foam in surfboards. Solid polyurethanes are use for inline skate wheels, tables and furniture, and for encasing electronic components. Polyurethanes are also the bases of some varnishes.

Acrylonitrile-butadiene – Styrene Copolymer – ABS

ABS combines the strength and stiffness of acrylonitrile and styrene with the flexibility of butadiene rubber. It is used for computer keyboards. Its toughness and impact resistance are used in canoe hulls.

Epoxies

An epoxy is a thermosetting resin that polymerizes when mixed with a catalyst or hardener. Most common resins are produced from a reaction between epichlorohydrin and bisphenol-A. Epoxy materials tend to harden somewhat more gradually than polyesters. Epoxies are widely used as adhesives.

Silicones

Silicones are analogous to linear polymers. They have an S-O backbone with organic radicals or hydrogen attached to the silicon atoms, as illustrated in Figure 11.23, with organic radicals or hydrogen attached to the silicon atoms. Silicones are stable up to 250 °C and are resistant to sunlight, oxidation, and chemical attack. They have low toxicity.

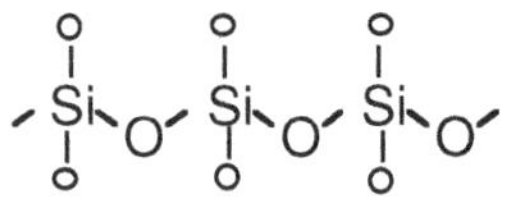

Figure 11.23. Structure of silicones. The small open circles represent either CH_3 groups or other hydrocarbons.

The uses of silicones include body implants, sealants for joints in construction, O-rings, and spatulas for cooking. Silicone greases for lubrication withstand higher temperatures than hydrocarbons.

Notes of Interest

Nylon was first synthesized by Wallace Carothers at DuPont in February 1935. Its first commercial use was in 1938 as bristles for toothbrushes. In 1940 it was used to replace silk in women's stockings. During World War II the entire production of nylon was used for parachutes, ropes, flak vests, vehicle tires, combat uniforms, and many other military items.

Bakelite was the first thermosetting plastic. It was invented by Belgian Leo Baekeland during 1907 to 1909. It is generally filled with wood flour. It quickly found use as an electrical insulator because of its electrical resistance and capability to withstand moderately high temperatures. Today it has largely been replaced by urea formaldehyde.

Problems

1. Calculate the average molecular weight of polypropylene having a polymerization degree of 260.
2. For the reaction to make nylon (Figure 11.15) how many pounds of water are released in each pound of nylon produced?
3. Discuss the recycling capability of thermoplastic and thermosetting polymers.
4. Calculate the weight percent of sulfur that would cause half of the double bonds in polyisoprene to be cross-linked with S-S.

12   Polymer Processing

Cooling

Polymerization often releases a large amount of heat. Unless the heat is removed the reactants will become too hot.

Example Problem 12–1:

a. How much energy is released when one mole of phenol reacts with one mole of formaldehyde?
b. If the process were adiabatic (no heat released to the surroundings), how much would the temperature rise? The heat capacity of phenol formaldehyde is 1.193 kJ/kg °C.

Solution:

a. Using the data in Appendix 7, the energy to break a C-O bond and an N-H bond is 360 and 430 kJ/mole, and the energy released in forming a C-N bond and an O-H bond is 305 and 500 kJ/mole. The net energy release is 805 $-360 =$ 15 kJ/mole.
b. The molecular weight of phenol formaldehyde is 90 g/mole ($2O = 32 + 2N = 28 + 2C = 24 + 6H = 6$) and the molecular weight of water is 18: $\Delta T = 15$ kJ/ $[(0.090 \text{kg}) \, (1.193 \, \text{kJ/kg} °\text{C}) + \, 0.018 \text{kg})(4.186 \, \text{kJ/kg} °\text{C})] \; = \; 82 °\text{C}$. This would require cooling.

Injection Molding

Injection molding is similar to die casting of metals. A molten thermoplastic is injected into a metal mold at high pressure, and the molded part is ejected after it cools sufficiently. Injection molding is used to make a wide variety of parts, from small components to entire auto body panels. The most commonly used thermoplastics are polystyrene, ABS, nylon, polypropylene, polyethylene, and PVC. Although the properties of finished products benefit from a high molecular weight, very high molecular weights are not desirable for injection molding because viscosity increases with molecular weight. If the molecular weight is too high, injection molding may be impossible or impractical.

Compression Molding

Compression molding is a widely used molding process for thermosetting plastics. The process consists of loading a precise charge in the bottom half of a heated mold, closing the mold halves to compress the charge, forcing it to flow and conform to the shape of the cavity, heating the charge to polymerize it into a solidified part, and finally removing the part from the cavity (Figure 12.1). Materials for compression molding of thermosetting plastics include phenolics, melamine, urea formaldehyde, epoxies, urethanes, and elastomers. Typical products include electric plugs, sockets, housings, pot handles, and dinnerware plates.

Applications of compression molding of thermoplastics include phonograph records and rubber tires. For both thermosetting and thermoplastic polymers, the initial charge of molding compound can be in the form of a powder or pellet, liquid, or a preform. The amount of polymer must be precisely controlled for consistency in the molded product. Often the charge is preheated to soften the polymer and shorten the production cycle time.

Molds for compression molding are generally simpler than injection molds. There are no sprues or runner systems in a compression mold. The process is generally limited to simple part geometries. The advantages of compression molding over injection molding include less expensive molds that require low maintenance, less scrap, and lower residual stresses.

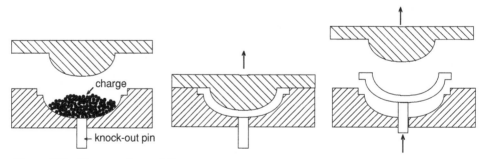

Figure 12.1. Compression molding.

Transfer Molding

In this process, a charge of a thermosetting polymer is loaded into a heated chamber to soften before it is placed into a heated mold where polymerization occurs. Scrap or cull from each cycle in the form of the leftover material in the base of the well and

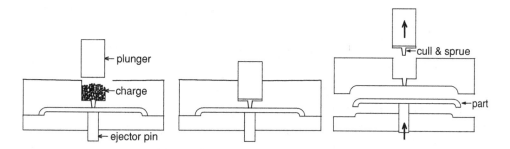

Figure 12.2. Transfer molding of a dish.

lateral channels is discarded because its thermosetting scrap cannot be recovered (Figure 12.2).

Transfer molding is closely related to compression molding because it is utilized on the same polymer types (thermosets and elastomers). It is similar to injection molding in that the charge is preheated in a separate chamber and then injected into the mold.

Bulk Extrusion

In the extrusion of thermoplastics, raw material (often called *resin* in the industry) is in the form of small beads that are gravity fed from a hopper into the barrel of the extruder. Additives such as colorants and ultraviolet inhibitors (in either liquid or pellet form) are often mixed into the resin before it is fed into the hopper. The material enters through the feed throat (an opening near the rear of the barrel) to a rotating screw that forces it forward into the barrel. As the thermoplastic is pushed through the barrel, it is heated and gradually melts (usually around 200 °C) as illustrated in Figure 12.3. Additional heat is generated by the deformation and friction in the barrel so if an extrusion line is running fast enough, sometimes the heaters can be turned off. The molten plastic is forced through a die that gives the final product its shape. The product is cooled by pulling the extruded shape through a water bath. Because plastics are very good thermal insulators, they are difficult to cool quickly. To keep tubes and pipes from collapsing, the water bath is under a

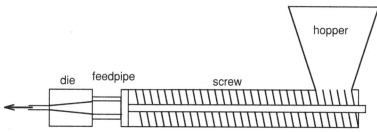

Figure 12.3. The components of a plastic extruder.

controlled vacuum. With plastic sheeting, the cooling is achieved by pulling through a set of cooling rolls. Once the product has cooled, it can be spooled or cut into lengths for later use. Extrusion is used to produce rigid PVC pipes and window frames. Plasticized PVC is extruded around wires to form the insulation.

Blown Films and Blow Molding

Films to form bags and sheets are made by blowing air into a tube that has been extruded. The air expands the tube, which is then collapsed by nip rollers. The thickness and size of the film is controlled by the volume of air inside the bubble, the speed of the nip rollers, and the extruders' output rate.

Plastic bottles are made by blow molding. The process begins with the conventional extrusion of a *parison* or closed-end tube. The parison is then clamped inside a hollow mold and inflated. The air pressure forces the parison against the mold surface until it cools in the shape of the interior of the mold cavity. The mold is then opened and the plastic bottle is ejected. Blow molding is used with many different plastics, including HDPE, PVC, PC, PP, and PET. The largest use is in PET beverage bottles. The same process of blow molding is used to make glass containers.

Fiber Drawing and Rolling

When fibers of thermoplastics are stretched, the molecules become aligned and the fibers become very strong and resist being stretched any further. Once the fibers have been stretched, or *drawn*, they are strong enough for use in textiles and rope. Drawn fibers of nylon and polyesters are used for their strength.

Sheets of polymethylmethacrylate, polycarbonate, polyethylene, polystyrene, nylon, acrylics, and other polymers are produced by rolling. The process is similar to the rolling of metals.

Thermoforming

Sheets of thermoplastics, heated enough to soften them, are formed into shapes by drawing them into a mold with vacuum as illustrated in Figure 12.4.

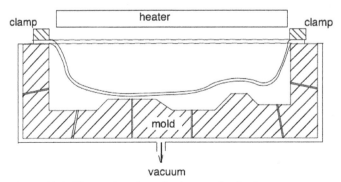

Figure 12.4. Thermoforming of a thermoplastic sheet.

Foaming

Foams may be created mechanically by a chemical reaction that forms a gas, or by heating a polymer in which a gas is dissolved. Polystyrene foam is formed by the release of dissolved pentane on heating. Polyurethane is foamed by the release of water vapor produced as a by-product of the polymerization reaction. In all cases the foam is formed into the desired shape by molding.

Notes of Interest

Many consumer goods are shrink-wrapped. Shrink-wrapping uses PVC or low-density polyethylene films that have been stretched biaxially during manufacture. On heating they tend to contract so if wrapped around an object and heated with hot air, they shrink to form a tight package.

The same phenomenon is used to make tubes that serve as electrical insulators. Prestretched tubes placed over a wire joint decrease their diameters, forming a tight insulation when heated.

Michael J. Owens developed the first automatic glass-blowing machine for making bottles. Born in 1878 he joined the new glass company founded by Edward Libby in 1888. With Libby's financing, he developed machinery to replace the slow one-at-a-time blowing of bottles by individuals in 1903. The cheap bottles made possible by this development allowed bottles to be used for ordinary food and beverages rather than just luxury items. His development led to the formation of the Owens bottle company, which later merged with the Illinois Glass Company to form Owens-Illinois. Owens also developed mass production techniques for window glass and did early research on glass fibers.

Problems

1. Explain why very-high-molecular-weight thermoplastics are hard to process into useful shapes.
2. Explain why the glass transition temperature of a copolymer is lower than the average of two polymers.
3. Calculate the amount of heat per kilogram that must be removed during the polymerization of Nylon 6/6. See Appendix 6 for bond energies.
4. Discuss the difference in mold temperature for compression molding of thermosetting vs. thermoplastic resins.
5. What are the ecological advantages of thermoplastics over thermosets?

13 Glasses

Glasses are amorphous so they have no long-range order and no symmetry. There is, however, a great deal of short-range order. If crystallization is prevented during cooling, an amorphous glass will form with short-range order inherited from the liquid. The critical cooling rate to prevent crystallization varies greatly from one material to another. Silicate glasses cannot crystallize unless the cooling rates are extremely slow. On the other hand, extremely rapid cooling is required to prevent crystallization of metals.

Structure of Silicate Glasses

The basic structural units of silicate glasses are tetrahedra with Si^{4+} ions in the center bonded covalently to O^{2-} ions at each corner. In pure silica all corner oxygen ions are shared by two tetrahedra (Figure 13.1). The result is a covalently bonded glass with a very high viscosity at elevated temperatures.

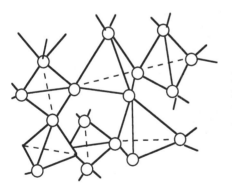

Figure 13.1. Silica glass is composed of tetrahedra with four O^{2-} ions surrounding Si^{4+} ions at the center. Each O^{2-} ion is shared by two tetrahedra. From W. F. Hosford, *Materials Science: An Intermediate Text*, Cambridge, 2006.

The compositions of typical commercial glasses are quite complex. Soda-lime glasses may contain 72% SiO_2, 14% Na_2O, 11% CaO, and 3% MgO. The Na^+, Ca^{2+}, and Mg^{2+} ions are bonded ionically to some of the corner O^{2-} ions (Figure 13.2). With these alkali and alkaline earth oxides, not all of the oxygen ions are covalently bonded to two tetrahedra. This lowers the viscosity at high temperatures.

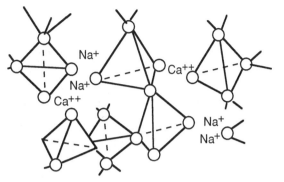

Figure 13.2. Commercial glasses contain alkali and alkaline earth ions, which substitute ionic bonds for the covalent bonds between tetrahedra. From W. F. Hosford, *ibid*.

Zachariasen* described four requirements for formation of glass from oxides:

1. Oxygen atoms are linked to no more than two anions.
2. The coordination number of the glass-forming anion is small (three or four).
3. The polyhedra are formed by oxygen ions sharing corners, not edges or faces.
4. Polyhedra are linked in a three-dimensional network.

The only oxides to satisfy these requirements are A_2O_3, AO_2, and A_2O_5, where A is an anion. Triangular coordination with A_2O_3 and tetrahedral coordination with AO_2 and A_2O_5 are possible.

Chemical Compositions

The chemical components of silicate glasses can be divided into three groups:

1. *Glass formers* include SiO_2 and B_2O_3. In a pure B_2O_3 glass the boron ions are in the center of a circle surrounded by three oxygen ions, each of which is shared with another triangle. This is possible because of the very small size of the B^{3+} ion.
2. *Modifiers* are alkali and alkaline earth oxides such as Na_2O, K_2O, CaO, and MgO. The Na^+, K^+, Ca^{2+}, and Mg^{2+} ions are bonded ionically to the oxygen ions at the corners of the silica tetrahedra. This results in *non-bridging* oxygen ions. Modifiers tend to decrease the overall bond strength and thereby lower the viscosity.
3. *Intermediates* such as Al_2O_3 and PbO do not form glasses themselves, but may join in the silica network. When Al_2O_3 is added to glass, some of the Al^{3+} ions act as intermediates, occupying centers of tetrahedra, and some act as modifiers.

Bridging vs. Non-bridging Oxygen Ions

In a silicate glass, each monovalent cation (Na^+ or K^+) causes one unbonded O^{2-} corner, and each divalent cation (Ca^{2+} or Mg^{2+}) causes two unbonded O^{2-} corners. The properties change as the number of unbonded corners increases. The number

---

* W. H. Zachariasen, *J. Amer. Chem Soc.*, v. 54, 1932, p. 3841.

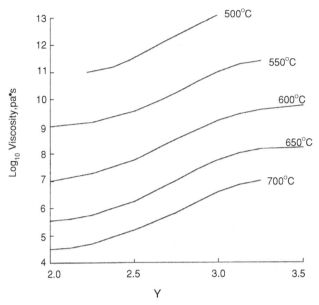

Figure 13.3. The viscosity of silicate glasses drops with decreasing O/Si ratios. The data from H. J. L. Trap and J. M. Stevels, *Gtech. Ber*, v. 6, 1959, are for equal molar additions of Na_2O, K_2O, CaO, SrO, and BaO. From W. F. Hosford, *Materials Science: An Intermediate Text*, Cambridge, 2006.

of unbonded corners per tetrahedron is $(2D + M)/S$ where D is the mole fraction of divalent modifiers, M is the mole fraction of monovalent modifiers, and S is the mole fraction of SiO_2. The number of bridging oxygen ions per tetrahedron, Y, is

$$Y = 4 - (2D + M)/S. \qquad 13.1$$

Glass Viscosity

The viscosities of glasses increase with the fraction of bridging oxygens. Figure 13.3 shows the increase of η with Y.

Example Problem 13–1:

Predict the viscosity at 700 °C of a glass containing 16 mole % CaO, 14 mole % Na_2O, and 4 mole % MgO with the rest being SiO_2.

Solution: Using equation 12.1, $Y = 4 - (2D + M)/S = 4 - [2(16 + 4) + 14]/66 = 3.18$. From Figure 13.3, $\log_{10}\eta = 6.5$ $\eta = 10^{6.5} = 3.2$ MPa-s.

Figure 13.4 is a plot of the viscosities of several glass compositions as a function of temperature. Note that, above the glass transition temperatures, the temperature dependence of the viscosity can be described by an Arrhenius equation,

$$\eta = A\exp[-Q/(RT)]. \qquad 13.2$$

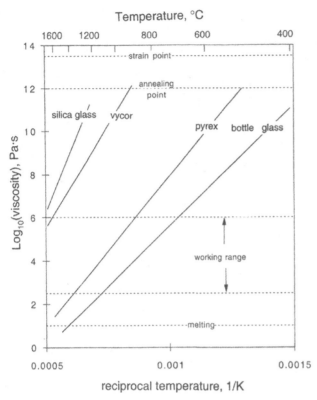

Figure 13.4. Temperature dependence of viscosity for several glasses. The working range is the temperature range in which glasses can be economically shaped. The straight lines on the semi-log plot do not extend below the glass transition temperature.

Below their glass transition temperatures, however, glasses should be regarded as solids rather than supercooled liquids. The viscosity increases much more rapidly with decreasing temperature than what is indicated by equation 13.2.

Several temperatures are identified in terms of viscosity. In the *working range* (10^3 to 10^7 Pa-s) glass can be shaped economically. The *softening point* is defined by 4×10^6 Pa-s). Above this temperature the weight of a glass object will cause appreciable creep. Stress relief occurs in the annealing range ($10^{11.5}$ to $10^{12.5}$ Pa-s).

Thermally Induced Stresses

Increased contents of alkali and alkaline earth oxides increase coefficients of thermal expansion in addition to lowering viscosities. Table 13.1 shows coefficients of thermal expansion of several glass compositions, and Figure 13.5 shows the relationship between the coefficient of thermal expansion and viscosity.

Glass, like most ceramics, is susceptible to fracturing under stresses caused by temperature gradients. Internal stresses in a material arise when there are different temperature changes in adjacent regions. Hooke's law becomes

$$\varepsilon_x = \alpha \Delta T + (1/E)[\sigma_x - \upsilon(\sigma_y + \sigma_z)] \qquad 13.3$$

Table 13.1. *Compositions and coefficients of thermal expansion of several glasses*

Glass	Composition, wt. %*	$\alpha \times 10^{-6}/°C$
Silica	100 SiO$_2$	0.5
Vycor	4 B$_2$O$_3$**	0.6
Pyrex	12 B$_2$O$_3$, 4 Na$_2$O, 4 Al$_2$O$_3$	2.7
Plate	13 CaO, 13 Na$_2$O, 2 MgO, 1 Al$_2$O$_3$	9.

*   In each case balance is SiO$_2$.
** The composition of finished Vycor. The composition before
    forming has more of other oxides.

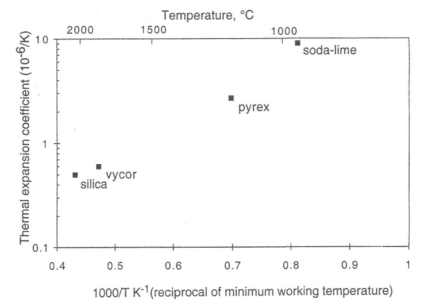

Figure 13.5. The relation between thermal expansion coefficient and the temperature at which the viscosity is 10^7 Pa-s. Compositions that promote lower working temperatures have higher coefficients of thermal expansion.

when temperature effects are added. Two regions, A and B, in intimate contact must undergo the same strains ($\varepsilon_{xA} = \varepsilon_{xB}$). If there is a temperature difference, $\Delta T = T_A - T_B$, between the two regions,

$$\alpha \Delta T + (1/E)[\sigma_{xA} - \sigma_{xB} + \upsilon(\sigma_{yA} + \sigma_{zA} - \sigma_{yB} - \sigma_{zB})] = 0. \qquad 13.4$$

Example Problem 13–2:

The temperature of the inside wall of a tube is $200\,°C$, and the outside wall temperature is $40\,°C$. Calculate the stresses at the outside of the wall if the tube is made from a glass having a coefficient of thermal expansion of $\alpha = 8 \times 10^{-6}/°C$, an elastic modulus of 10×10^6 psi, and a Poisson ratio of 0.3.

Solution: Let x, y, and z be the axial, hoop, and radial directions. The stress normal to the tube wall is $\sigma_z = 0$, and symmetry requires that $\sigma_y = \sigma_x$. Let the reference position be the mid-wall where $T = 120\,°C$. ΔT at the outside is $40\,°C - 120\,°C = -80\,°C$. The strains ε_x and ε_y must be zero relative to the mid-wall. Substituting in equation 19.9, $0 = \alpha\Delta T + (1/E)(\sigma_x + \upsilon(\sigma_y + \sigma_z))$, $\alpha\Delta T + (1-\upsilon)\sigma_x/E = 0$, so $\sigma_x = \alpha E\Delta T/(1-\upsilon)$. $\sigma_x = (8 \times 10^{-6}/°C)(80\,°C)(10 \times 10^6 \text{ psi})/0.7 = 9{,}140$ psi.

In general, the stresses caused by temperature differences are proportional to α, $E/(1-\upsilon)$, and ΔT. The parameter

$$R_1 = \sigma_f(1 - \upsilon)/(E\alpha) \qquad 13.5$$

describes the relative susceptibility to thermal shock. A different thermal shock parameter,

$$R_2 = K_{Ic}/(E\alpha), \qquad 13.6$$

is based on the fracture toughness. If the length of pre-existing cracks is constant, these are equivalent since σ_f is proportional to K_{Ic}. Thermal conductivity has some influence on susceptibility to thermal shock because it influences the term ΔT. A high thermal conductivity lowers the value of ΔT. It should be noted that these parameters apply only to materials that are brittle. In materials that flow plastically, the stresses never rise above the stress to cause plastic flow so the right-hand side of equation 13.3 should be modified by the addition of a plastic strain term.

Because Young's modulus is nearly the same for all grades of glass, differences of thermal shock resistance are related to differences in thermal expansion. The compositions and thermal expansion coefficients of several glasses are listed in Table 13.1. The thermal shock resistance of silica glass and Vycor is much better than that of plate glass.

Vycor®

Vycor® was developed by Corning Glass to provide a way around the difficulty in forming glasses that have low coefficients of thermal expansion. The starting composition (62.7% SiO_2, 26.9% B_2O_3, 6.6% Na_2O, and 3.5% Al_2O_3) has a low enough viscosity at a reasonable temperature for the glass to be shaped. After it is shaped, the glass is heat treated between 500 and 750 °C. A spinodal reaction during this heat treatment separates the glass into two phases, one containing 96% SiO_2 and 4% B_2O_3, and a second impurity-rich phase. The impurity-rich phase is removed by acid etching leaving a silica-rich glass with about 28% porosity. This can either be used as a filter or reheated to allow sintering to produce a fully dense product.

Devitrification

If a glass is held for a long period at an elevated temperature it may start to crystallize or *devitrify*. Devitrification of fused quartz (silica glass) to the crystalline cristabolite is slow. Nucleation usually occurs at a free surface and is often stimulated by contamination from alkali ions such as sodium. The rate of growth of cristobolite is increased by oxygen and water vapor. With surface contamination, devitrification of fused quartz may occur at temperatures as low as 1000 °C. However, if the surface is clean it rarely occurs below 1150 °C.

Some glasses may be intentionally crystallized. David Stookey at Corning Glass discovered a way of producing fine-grained ceramics by crystallizing a glass. This process has been commercialized as Pyroceram® and Corningware®. It involves using a lithia-alumina silicate with TiO_2 as a nucleating agent. The material is first formed as a glass into its final shape at an elevated temperature and then heat treated at a lower temperature to allow nucleation of crystals. It is finally heated to a somewhat higher temperature to allow growth of the crystals. This process allows glass-forming processes to be used to obtain the final shape while producing a final product that is resistant to thermal shock because of a very low thermal expansion coefficient.

Photosensitive Glasses

Photosensitive glasses containing microcrystalline silver halides darken with exposure to ultraviolet light and lighten after the sunlight is removed. The darkening occurs when silver halides are reduced to metallic silver on exposure to ultraviolet.

Other Inorganic Glasses

Borax, B_2O_3, forms a glass in which the basic structural elements are triangles with boron at the center surrounded and covalently bonded to three oxygen atoms. Each of the oxygen atoms is shared by three triangles as shown in Figure 13.6.

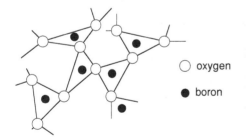

○ oxygen

● boron

Figure 13.6. Borax glass. Each boron atom is covalently bonded to three oxygen atoms, which form a triangle around the boron atom. Each oxygen atom is shared by two triangles. From W. F. Hosford, *Materials Science: An Intermediate Text*, Cambridge, 2006.

Chalcogenide* glasses consist of long Se (or Te) chains bonded with Ge or As. The basic structural units are chains that are cross-linked by As or Ge. The structure of molten pure Se and pure Te consists of long chain molecules. These form glasses

---

* Chalcogens are O, S, Se, and Te.

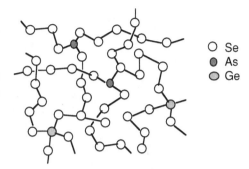

Figure 13.7. Schematic of a chalcogenide glass. From
W. F. Hosford, *Materials Science: An Intermediate Text*,
Cambridge, 2006.

if cooled rapidly; however, they will crystallize if heated between the glass transformation temperature and the melting point. Small amounts of As or Ge will form a network and prevent crystallization. Figure 13.7 shows the structure schematically. Such glasses are useful in xerography.

Residual Stresses

The fracture resistance of glass can be improved by inducing a pattern of residual stress with the surface under compression. The surface compression must, of course, be balanced by a residual tension in the center. Residual compression in the surface increases fracture resistance because fractures almost always start at the surface, and the stresses are highest at the surface if they involve any applied bending or torsion. Furthermore, defects are much more likely to be present at the surface. With residual compression at the surface, greater external loads can be tolerated without fracture.

Compressive residual stresses can be induced in the surface by either cooling rapidly from high temperature or by chemical treatment. In the former process, called *tempering*, the glass is cooled with jets of air. During the cooling, the surface undergoes a thermal contraction before the interior. The different amounts of thermal contraction are accommodated by viscous flow of the interior. When the interior finally cools, dimensional compatibility is maintained by elastic contraction (compression) of the surface. Glass may also be *chemically tempered* by ion exchange. Glass is immersed in a molten salt bath containing potassium ions. Some of the K^+ ions diffuse into the glass, replacing Na^+ ions. Since the K^+ ions are larger than the Na^+ ions they replace, the region affected is left in compression. One important difference between the two processes is the depth of the compressive layer. The depth of the region under compression is much less in the chemically tempered glass. Not only is tempered glass more resistant to fracture than untempered glass, but it also breaks into much smaller pieces, which are less dangerous than the large shards produced in fracture of untempered glass. Tempered glass is used for the side and rear windows of automobiles for these reasons. The windshields of automobiles are made from safety glass produced by laminating two pieces of glass with a polymeric material (polyvinyl butyral) that keeps the broken shards from causing injury. Figure 13.8 shows fracture patterns typical of untempered glass, laminated safety glass, and tempered glass.

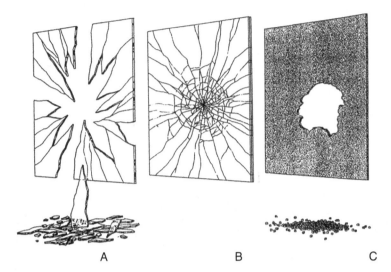

Figure 13.8. Typical fracture patterns of three grades of glass: (A) annealed, (B) laminated and (C) tempered. From *Engineered Materials Handbook v. 4, Ceramics and Glasses*, ASM, 1991.

Strength of Glasses

The tensile strengths of glass and other brittle materials are much more sensitive to surface flaws than more ductile materials because stress concentrations cannot be relieved by deformation. Therefore, strengths of glass are quite variable. Typical strengths of various grades of glass are listed in Table 13.2, but there is considerable scatter from these values.

Shaping Glass

Bottles are made by the same blowing process described in Chapter 12 for plastic bottles. Plate glass is formed by floating molten glass on the surface of molten tin, which produces a very smooth surface. Glass fibers are made by extruding molten glass through very small holes in a *spinerette*. The glass fibers must be prevented from contacting each other before they are coated with a *size* (polymer coating weighing

Table 13.2. *Typical fracture strengths of several glasses*

Type of glass	Strength (MPa)
Pristine fibers	7,000
Typical fibers (E-glass)	3,500
Float glass (windows)	70
Tempered window glass	150
Container glass	70
Fused silica	60

only about 1% of the fiber weight). They would otherwise abrade each other, causing a severe loss of strength.

Metallic Glasses

Crystallization can be prevented in certain alloys if they are cooled rapidly enough. Metallic glasses will form in these cases. Compositions of glass-forming alloys have several common features:

1. The equilibrium diagrams consist of two or more phases. Redistribution of the elements by diffusion is necessary for crystallization.
2. The compositions correspond to deep wells in the equilibrium diagram so the liquid phase is stable at low temperatures where diffusion is slow. See Figure 13.9.
3. The compositions usually have large amounts of small metalloid atoms like B, C, P, Be, and Si.

The formation of metal glasses by rapid cooling was first reported by Paul Duwez and coworkers in the 1960s.* They achieved cooling rates of thousands of degrees per second by shooting a fine stream of liquid metal onto a water-cooled copper drum. With the early compositions, cooling rates of about 10^5 °C/s were necessary to prevent crystallization. This limited alloys to thin ribbons or wires.

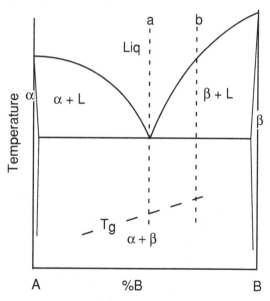

Figure 13.9. Schematic phase diagram. Composition a is more likely to form a glass than composition b because of the much lower temperature at which crystallization can start. From W. F. Hosford, *Materials Science: An Intermediate Text*, Cambridge, 2006.

More recently magnesium-base, iron-base, and zirconium-titanium-base alloys have been developed that do not require such rapid cooling. In 1992, W. L. Johnson and coworkers developed the first commercial alloy available in bulk form: Vitreloy 1,

* W. Klement, R. H. Willens, and P. Duwez, *Nature*, v. 187, 1960, p. 869.

which contains 41.2 a/o Zr, 13.8 a/o Ti, 12.5 a/o Cu, 10 a/o Ni, and 22.5 a/o Be. The critical cooling rate for this alloy is about 1°C/s so glassy parts can be made with dimensions of several centimeters; properties are given in Table 13.3.

Table 13.3. *Properties of Vitreloy 1*

Density	5.9 mg/m^3
Young's modulus	95 GPa
Shear modulus	35 GPa
YS	1.9 GPa
K_{IC}	55 MPa$\sqrt{m}$
T_g	625 K
Endurance limit/UTS	≈ 0.03

Metal glasses have very high yield strengths that permit very high elastic strains and the storage of a large amount of elastic energy. The high ratio of yield strength to Young's modulus permits elastic strains of $1.9/95 = 2\%$. The tensile stress-strain curves of glassy metals show almost no work hardening. Tensile tests are characterized by serrated stress-strain curves resulting from sudden bursts of deformation localized in narrow shear bands with an abrupt load drop. The net effect is that the total plastic strain is quite limited. The localization can be explained partially by the lack of work hardening. The formation of free volume and adiabatic heating have been offered as explanations. The fracture toughness is very high, but the fatigue strength is very low. The ratio of endurance limit to yield strength of 0.03 is much lower than the ratios of 0.3 to 0.5 typical of crystalline metals. Metallic glasses have good corrosion resistance and low damping.

There are two principle uses for glassy metals. Because metal glasses have no barriers for domain wall movement they are excellent soft magnetic materials. Thin ribbons have been used for transformer cores since the 1960s.

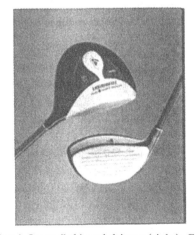

Figure 13.10. Golf club heads of Vitreloy 1. Irons (left) and drivers (right). Courtesy of Otis Buchanan, Liquidmetal Technologies, Lake Forest, California.

The other major application is based on the large amount of elastic energy that can be stored. The very high yield strengths typical of metallic glasses permit very high elastic strains and therefore storage of a large amount of elastic energy. Commercial use has been made of metallic glass in the heads of golf clubs (Figure 13.10). The great capacity to store elastic energy has permitted longer drives.

Notes of Interest

According to the Roman author Pliny the Elder, the Phoenicians made the first glass. He stated that sailors on a ship loaded with ammonium nitrate used it to support their pots while cooking a shore meal. It was melted by the heat of the fire and mixed with sand to form glass.

It is often taught that the stained glass windows in European cathedrals are thicker at the bottom than at the top because of creep during the centuries; however, glass does not creep at ambient temperatures. The reason that the glass is thicker near the bottom than at the top is because it was installed that way.

Problems

1. What fraction of the oxygen ions in a glass that has a composition of 70 mole % silica, 15% mole Na_2O, 11% mole CaO, and 4% mole MgO are covalently bonded?
2. The viscosity of leaded glass extrapolated to 200 °C is 10^{20} Pa-s. The stress at the bottom of a piece of glass 1 m high from its own weight is about 25 kPa. Calculate the strain that would accumulate after five centuries. Make an unreasonable assumption that the temperature was 200 °C! Note the strain rate, $\dot{\varepsilon} = \sigma/(\sqrt{3}\eta)$, where η is the viscosity and σ is the stress.
3. Calculate the linear shrinkage that accompanies the sintering of Vycor® to full density.
4. Why aren't metals with high values of $E\alpha[\sigma_f(1-\upsilon)]$ subject to thermal shock?
5. Calculate the amount of elastic energy per volume that can be stored in Vitreloy 1.

14 Crystalline Ceramics

Ceramics

Ceramics are compounds consisting of metal and non-metal ions bonded either covalently or ionically. Most ceramics are crystalline. They tend to have high melting points and be very hard and brittle. Their tensile strengths are limited by brittle fracture but their compressive strengths are high. Because they retain high hardnesses at elevated temperatures, they are useful as *refractories** such as furnace linings. Oxidation at high temperature is not a problem with refractory oxides as it is with refractory metals. Magnesia, alumina, and silica are used for furnace linings. Ceramics are also used as tools for high-speed machining of metals. The high hardness of ceramics at room temperature leads to their use as abrasives as either loose powder or bonded into grinding tools. The low ductility of ceramics limits the structural use of ceramics mainly to applications in which the loading is primarily compressive. Iron-containing ceramic are used as magnets.

The bonding strength depends on the valences of the metal and non-metal. Compounds with higher valences (e.g., SiC, Si_3N_4) tend to be more strongly bonded than those of lower valences (e.g., $NaCl$, MgO) so they have higher melting points and higher hardnesses. Appendix 9 covers the geometric principles governing the crystal structures adopted by various ceramic compounds.

Slip Systems

Ceramics tend to be brittle because they have a limited number of independent slip systems. In an ionically bonded crystal, the slip plane and direction must be such that there is not close contact between ions of like sign as slip occurs. This is illustrated in Figure 14.1. For some crystal structures there are more than one type of slip system, each with a different stress required to operate.

---

* The word *refractory* means resistant to high temperatures.

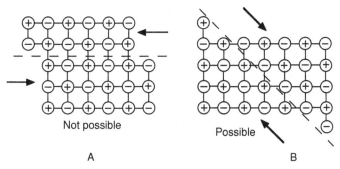

Figure 14.1. Schematic illustration of the principle that the slip systems are such that, on translation, like-sign ions do not come in contact. In (A), the mutual repulsion between ions of like sign during slip would cause failure. From W. F. Hosford, *Mechanical Behavior of Materials*, Cambridge, 2005.

Ionic Bonding

Ionic bonds are the result of the mutual attraction of ions of opposite charge. The energy of the bond, U_{pair}, between a pair of oppositely charged ions depends on the charges on the ions and their separation,

$$U_{\text{pair}} = -z_1 z_2 e^2/d, \hspace{3cm} 14.1$$

where z_1 and z_2 are the valences of the ions, e is the charge on the electron, and d is the separation between ions. For a crystal, the bonding energy for an ion is the sum of the attractions and repulsions between all of the other ions in the crystal. This can be expressed as

$$U = -M z_1 z_2 e^2/d, \hspace{3cm} 14.2$$

where M is a constant between 1.5 and 2 that depends on the crystal structure.

Melting points and elastic moduli depend on the strength of bonds. Since $U = -M z_1 z_2 e^2/d$, the binding energies of ionic crystals of formula AB increase with the square of z. The melting points of ionic solids correlate very well with z^2/d (Figure 14.2). The melting points of oxides and carbides are listed in Table 14.1.

Figure 14.3 shows that the elastic constant for AB compounds is approximately proportional to z^2/d^4.

Hardness

Hardness also increases with z/d. The high hardness of ceramics is the basis for their use as cutting tools and as abrasives for grinding. Aluminum oxide cutting tools are used for high-speed machining because they retain their high hardness even when very hot. Grinding wheels are frequently made of SiC or Al_2O_3. Powdered silicon carbide, silica, and aluminum oxides are often used for polishing. Table 14.2 gives the hardness of some ceramics.

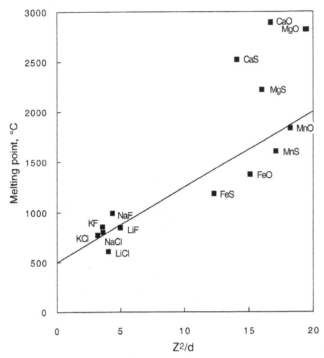

Figure 14.2. The melting points of AB ionic crystals increase with z^2/d. From W. F. Hosford, *Materials Science: An Intermediate Text*, Cambridge, 2006.

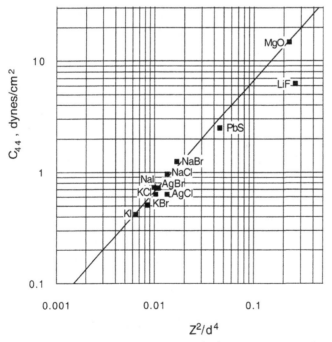

Figure 14.3. The correlation between z^2/d^4 and the elastic modulus, c_{44}, for various AB compounds with an NaCl structure. Data from J. J. Gilman in *Progress in Ceramic Science*, v. 1, Pergamon Press, 1961. From W. F. Hosford, *Materials Science: An Intermediate Text*, Cambridge, 2006.

Table 14.1. *Melting points of several oxides and carbides*

Compound	Melting point (°C)	Compound	Melting point (°C)
Al_2O_3	2050	Cr_3C_2	1890
BeO	2530	Cu_2O	1235
Cr_2O_3	1990	MoC	2692
Cu_2O	1235	SiC	2600
FeO	1420	TaC	3880
Fe_2O_3	1565	VC	2810
MgO	2800	WC	2870
MnO	1650	ZrC	3540
NiO	1990		
SiO_2	1710 (<1470 quartz)	B_4C	2450
TiO_2	1640	BN	(sublimes at 3000)
ZrO_2	2980		

Table 14.2. *Hardness of several ceramics*

Material	Vickers hardness (kg/mm^2)
Soda-lime glass	600
Fused silica	650
Si_3N_4	1700
Al_2O_3	1750
SiC	2600
B_4C	3200
W_2C	2400

Porosity

Most ceramics are prepared from powder by pressing the powder to the desired shape and then sintering at a high temperature (see Chapter 15). This process results in some degree of porosity, which causes some loss of strength as shown in Figure 14.4. This dependence can be approximated by the equation

$$\sigma = \sigma_o \exp(-bP), \qquad\qquad 14.3$$

where P is the volume fraction of pores.

The value of b in equation 14.3 is typically near 7 as shown in Example Problem 14-1.

Example Problem 14–1:

Determine the constant b in equation 14.1 that fits the data in Figure 14.4.

Solution: From equation 14.1, $\sigma_2/\sigma_1 = \exp[-b(P_2-P_1)$ or $b = -\ln(\sigma_2/\sigma_1)/(P_2-P_1)$. Substituting $\sigma_1 = 0.23$ at $P_1 = 0.2$ and $\sigma_2 = 0.46$ at $P_2 = 0.1$, $b = 7$.

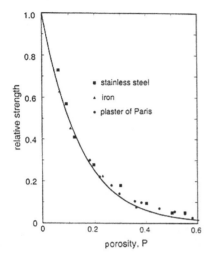

Figure 14.4. Decrease of fracture strength with porosity. From R. L. Coble and W. D. Kingery, *J. Am. Ceram. Soc.*, v. 29, 1956.

Porosity may be classified as either closed porosity or open porosity. Open pores are connected to the surface by channels so they can be filled with water.

Example Problem 14–2:

The true density of a ceramic is 3.71 Mg/m^3. A sample of sintered ceramic weighs 3.85 g when dry, 3.91 g when saturated with water, and 2.73 g when suspended in water. What is its true volume? What is its open porosity? What is its closed porosity?

Solution: The true volume is 3.85 g/(3.71g/cm^3) $= 1.038$ cm^3. The open porosity is $(3.91 - 3.85$ g)(1 cm^3 water/g)/1.038 cm^3 $= 0.0578$ or 5.78%.

The water displaced by immersion is $3.91 - 2.73 = 1.18$ g, so the total volume is 1.18 cm^3. The closed porosity is the total volume less the true volume less the open pore volume: $1.18 - 1.038 - 0.06 = 0.082$ or 8.2%.

Variability of Strength

The strengths of ceramics in tension are widely variable. The reason is that they fail in brittle fracture with strengths that depend on the lengths of internal cracks. This variability of strength is far greater than with metals, which usually fail in a ductile manner. Figure 14.5 illustrates this schematically. A statistical treatment of failure probability, called the Weibull analysis, is given in Appendix 10.

Although all ceramics are brittle, there are significant differences in toughness among them. Typical values of K_{Ic} given in Table 14.3 range from less than 1 to about 7 MPa$\sqrt{\text{m}}$.

If K_{Ic} is less than about 2 MPa$\sqrt{\text{m}}$, extreme care must be exercised in handling the ceramics. They will break if they fall on the floor. This limits severely the use of ceramics under tensile loading. Ceramics having K_{Ic} greater than about 4 MPa$\sqrt{\text{m}}$ are quite robust. For example, partially stabilized zirconia is used for metalworking tools.

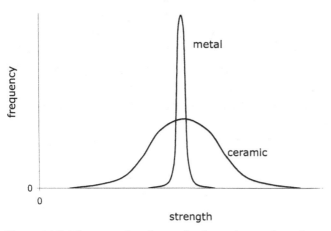

Figure 14.5. The strengths of ceramics depend strongly on internal defects. The variability of these defects differs because ceramics are brittle in contrast to metals, which are more ductile.

Toughening of Ceramics

Ceramics can be toughened by three basic energy-absorbing mechanisms: crack deflection, bridging cracks, and phase transformation.

Polycrystalline ceramics are usually tougher than single crystals of the same material because grain boundaries deflect cracks. The plane of a cleavage crack must change from grain to grain in a polycrystal; therefore, the crack is generally not normal to the tensile stress. The stress intensity factor, K, at the tip of a crack has a maximum value when the crack is normal to the tensile stress. This effect is evident in Table 14.3 when the toughnesses of single-crystal Al_2O_3 is compared with polycrystalline Al_2O_3. The toughnesses of crystallized glass is greater than that of typical glasses.

Table 14.3. *Fracture toughness of some ceramics*

Material	K_{Ic} (MPa$\sqrt{m}$)
Al_2O_3 (single crystal)	2.2
Al_2O_3 (polycrystal)	4
Mullite (fully dense)	2.0–4.0
ZrO_2 (cubic)	3–3.6
ZrO_2 (partially stabilized)	3–15
MgO	2.5
SiC (hot pressed)	3–6
TiC	3–6
WC	6–20
Silica (fused)	0.8
Soda-lime glass	0.82
Glass ceramics	2.5

Fibers, whiskers, or elongated grains of a second phase may form ligaments that bridge across an open crack and continue to carry some load (Figure 14.6). In this case shearing of the bonds between the matrix and these ligaments will absorb additional energy. This principle was used in ancient times to toughen clay bricks with straw or horsehair. Toughness can be increased by increasing the volume fraction of the reinforcing phase, weakening the bonds between the fibers and the matrix so energy is absorbed in pulling the fibers out of the matrix instead of breaking, or by using fibers of lower elastic modulus so they will not break before pullout. Figure 14.7 shows the toughening effect of SiC fibers in several ceramics.

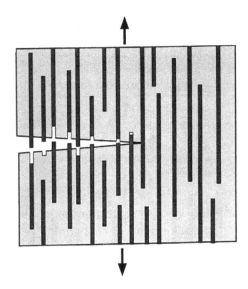

Figure 14.6. Schematic illustration of toughening by reinforcing ligaments. Shearing between the fibers and matrix during pullout absorbs additional energy. From W. F. Hosford, *Mechanical Behavior of Materials*, Cambridge, 2005.

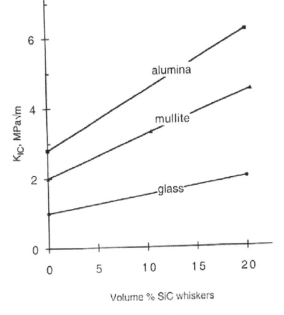

Figure 14.7. Toughening of several ceramics by addition of SiC fibers. Data from P. Becher, *J. Amer. Cer. Soc.*, v. 74. From W. F. Hosford, *ibid.*

A phase transformation in zirconia is used to increase its toughness. When 7% CaO or an appropriate amount of MgO is added to zirconia, its structure at high temperatures consists of two solid solutions: one cubic and one tetragonal. At low temperatures the tetragonal zirconia should transform to a monoclinic structure, but the transformation is too sluggish to occur. A sufficient stress at room temperature, however, will cause the transformation to occur by shear, with a volume expansion of about 4% and a shear strain of about 14%. The high stresses at the tip of an advancing crack trigger this reaction. Some of the energy applied externally is absorbed by the transforming particles, leaving less energy available for crack propagation. The volume expansion of the particles causes compressive stresses in the matrix, which shields the crack tip from the full stress intensity. This is illustrated schematically in Figure 14.8.

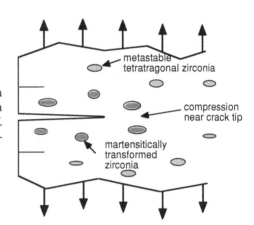

Figure 14.8. Stress-induced transformation of zirconia from tetragonal to monoclinic structures, creating a region of compression around the crack tip. From W. F. Hosford, *Mechanical Behavior of Materials*, Cambridge, 2005.

Growth of Single Crystals

Although most crystalline materials consist of millions of crystals, it is possible to grow single crystals. There are two main techniques: growth by gradually solidifying a liquid, and growth by precipitation from a supersaturated solution.

The Bridgman technique relies on directional solidification in a mold in such a way that there is a single nucleus. A mold containing molten material is slowly removed from a furnace so a crystal starting at one location grows directionally to form a crystal. A variant on this technique is the Czochralski method in which a seed crystal is lowered into a molten bath and then gradually raised and rotated as additional material freezes onto it. Figure 14.9 shows these two methods. The Czochralski method is used to produce large crystals of silicon for use as semiconductors.

Quartz crystals are grown from a solution of silica in water containing alkali metal salts, typically NaOH and $NaCO_3$ at 380 °C and a pressure of 1000 to 1500 bars or at 345 °C and 700 to 1,000 bars pressure. Quartz forms on a seed crystal at a slightly lower temperature with growth rates of about a half a millimeter per day. Ruby crystals are grown from solutions of alumina dissolved in molten salt. Single crystals of ceramics are used in lasers, oxygen sensors, and piezoelectric devices.

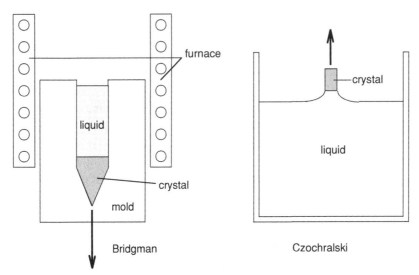

Figure 14.9. Bridgman and Czochralski methods of crystal growth.

Notes of Interest

Tungsten carbide has been used in armor-piercing ammunition because of its high density and hardness. The first use of W_2C projectiles was by the Luftwaffe in World War II against Russian T-34 tanks.

The name of a metal oxide is formed by substituting "a" for "ium" in the element's name. Thus aluminum oxide is *alumina*, magnesium oxide is *magnesia*, sodium oxide is *soda*, thorium oxide is *thoria*, and the oxide of silicon is *silica*.

The natural form of alumina is called corundum. Rubies and sapphires are single crystals of corundum with their characteristic colors caused by trace impurities.

Problems

1. Offer an explanation of why silicon carbide is harder than magnesium oxide.
2. Predict the loss of strength of MgO caused by 1% porosity.
3. How would having fibers bonded so long or so strongly bonded to the matrix that pullout cannot occur affect the toughness?
4. A brick that weighs 1.10 kg when dry weighs 1.15 kg when saturated with water and 0.53 kg when suspended in water.
   a. What is the open porosity?
   b. What is the bulk density?

15 Powder Processing

Many products are made by pressing and sintering powders. Most ceramics are consolidated by sintering. These include clay products as well as refractory oxides. These ceramics cannot be fabricated by melting and freezing. Sintering is also used to produce parts of metals that are difficult to melt. Examples include carbide tools and tungsten for lamp filaments. Mixed powders are sintered to make composites that are not otherwise possible such as friction materials for brakes and clutches. Porous parts for filters or oil-less bearings are made by incomplete sintering. Teflon cannot be melted without decomposing so it is also processed as a powder. Pharmaceutical pills are made from powder. Powder processing is a simple and cheap way of fabricating large numbers of parts.

Powder Compaction

Figure 15.1 illustrates schematically how a part is pressed from powder. The process is highly automated with many parts being pressed per second.

There are limitations on the shape of the die. The shape of the die must be prismatic so the compact can be ejected. The ratio of height to diameter must not be too great. Otherwise friction on the sidewalls of the die will not allow sufficient compaction pressure in the center, as illustrated in Figure 15.2. The loss of compacted density is greater as the ratio of height to diameter increases.

Sintering

Sintering pressed powders at elevated temperature bonds the small powder particles together without melting them. The driving force for sintering is the reduction in surface area and the associated energy. In the initial stages of sintering, adjacent particles adhere and form a neck where they are in contact. Figure 15.3 is a micrograph of such a neck formed between two nickel spheres.

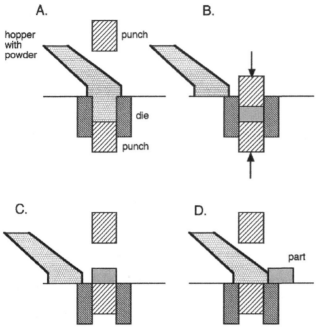

Figure 15.1. Powder compaction. A die is filled with powder (A), the powder is compacted (B), the compact is ejected (C), and the die is refilled as the part is pushed away (D).

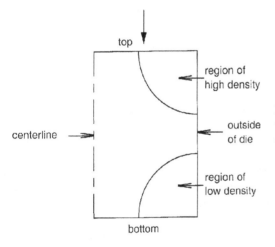

Figure 15.2. Sketch showing that friction along the sidewalls lowers the compaction pressure and compacted density in the center.

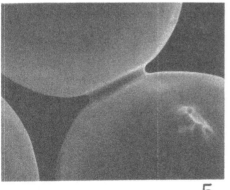

Figure 15.3. Partially sintered nickel spheres. The "5" is a micron marker. From R. M. German, *Powder Metallurgy Science*, Metal Powder Industries Federation, Princeton, N.J., 1984.

As sintering progresses, the necks between adjacent particles grow. There are two groups of sintering mechanisms, as shown in Figure 15.4. Some mechanisms, such as vapor transport and surface diffusion, take material from the surface to form the neck. These mechanisms do not change the distance between the centers of particles so they contribute little to densification. However, mechanisms that transport material from the interface between the particles to form the neck (grain boundary and lattice diffusion) do cause densification.

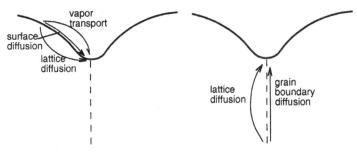

Figure 15.4. Growth of a neck by transport from the spherical surface (left) and from the grain boundary formed between the particles (right). From W. F. Hosford, *Materials Science: An Intermediate Text*, Cambridge, 2005.

Intermediate and Final Stages of Sintering

The unfilled edges between the particles form continuous pore paths as illustrated in Figure 15.5. These pores can be approximated as cylindrical tubes. The cylindrical pores become unstable when the ratio of their length to diameter exceeds π (i.e., $\ell \geq \pi d_p$). At this point, they collapse into more or less spherically shaped pores at the grain corners with diameters larger than the tubes. The collapse starts at about 15% total porosity and is complete by about 5% total porosity.

Pores remaining on the edges, faces, and grain interiors are isolated so the rate of shrinkage slows. Pores at corners may either shrink or grow, depending on the ratio of surface energies, γ_{SV}/γ_{SS}, where γ_{SV} is the vapor-solid surface energy and γ_{SS} is the solid-solid (grain boundary) surface energy and on the number of grains contacted by the pore (Figure 15.6).

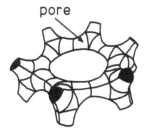

Figure 15.5. Continuous tubular pore structure. From R. M. German, *Powder Melallurgy Science*, Metal Powder Industries Federation, 1984, Fig. 6.11, p. 162.

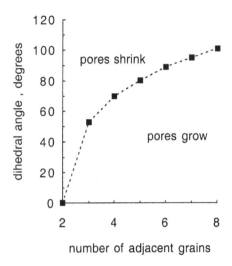

Figure 15.6. Whether a corner pore shrinks or grows depends on the dihedral angle and the number of grains it borders. Data from G. C. Kuczynski, in *Powder Metallurgy for High Performance Applications*, J. J. Burke and V. Weiss (Eds.), Syracuse Univ. Press, 1972. From W. F. Hosford, *Materials Science: An Intermediate Text,* Cambridge, 2006.

Loss of Surface Area

The surface area decreases throughout sintering. Figure 15.7 shows the decrease of surface area, S/S_o, of alumina as it was heated.

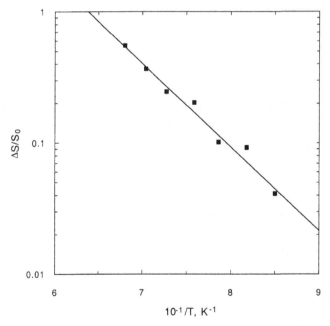

Figure 15.7. The loss of surface area as alumina is heated at 5 °C per minute. Data from S. H. Hillman and R. M. German, *J. Mat. Sci.*, v. 27, 1992. From W. F. Hosford, *ibid*.

Particle Size Effect

Of course, the time required for sintering is decreased as the particle size decreases. Figure 15.8 shows that if surface diffusion is the controlling mechanism, the time is inversely proportional to D^4.

The relation between two temperatures, T_1 and T_2, required for equal degrees of sintering of particles of diameters D_1 and D_2 is

$$1/T_2 = 1/T_1 - m(R/Q)\ln(D_2/D_1). \qquad 15.1$$

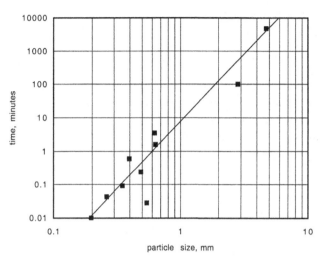

Figure 15.8. Time needed to reach a neck size of $X/D = 0.1$ in the sintering of various-size ice particles. The controlling mechanism is surface diffusion. Data from W. D. Kingery, *J. Appl. Phys.*, v. 31, 1960, pp. 833–838. From W. F. Hosford, *Materials Science: An Intermediate Text*, Cambridge, 2006.

Grain Growth

Grain growth occurs during the later stages of sintering. Figure 15.9 shows data for grain growth in alumina at $1550\,°C$.

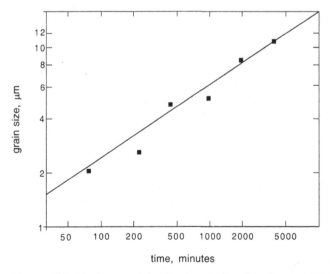

Figure 15.9. Grain growth in alumina during sintering at $1550\,°C$. Data from R. L. Coble, *J. Appl. Phys.*, v. 36, 1965. From W. F. Hosford, *Materials Science: An Intermediate Text*, Cambridge, 2006.

Once grain growth has allowed the grain boundaries to leave the pores, densification is slowed because the only mechanism of densification is by lattice diffusion of atoms from grain boundaries or dislocations to the pores. It should be noted that lattice diffusion also allows large isolated pores to grow at the expense of smaller pores because of the higher pressure inside the smaller ones.

Activated Sintering

Sometimes the addition of a very small amount of a second material greatly increases the rate of sintering. Usually this can be attributed to the formation of a phase with a much lower melting point in which the diffusion is much faster. Figure 15.10 shows that the sintering rate of tungsten is drastically increased by certain elements that form a very thin layer at grain boundaries.

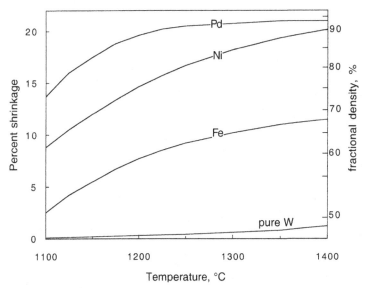

Figure 15.10. Sintering of tungsten is activated by addition of small amounts of Pd, Ni, Fe, or other elements. Data are for one hour at the indicated temperatures. Data from R. M. German, *Powder Metallurgy Science*, Metal Powder Industries Federation, 1984. From W. F. Hosford, *Materials Science: An Intermediate Text*, Cambridge, 2006.

Example Problem 15–1:

Experiments have shown that, under a set of sintering conditions, a powder compact with 82% density sinters to 98% density. A circular disc 2.30 cm in diameter and 1.5 cm thick is to be produced. What should be the diameter of the die for pressing the compact?

Solution: The ratio of the volumes after sintering to that before sintering is 0.82/0.98 = 0.837. The ratio of the initial linear dimensions before sintering to those after sintering is $[0.98/0.82]^{1/3} = 1.061$, so the die diameter should be $1.061(2.30) = 2.44$ cm.

Example Problem 15–2:

Experiments on sintering of stainless steel yielded the following results:

Temperature (°C)	Time (hr)	Shrinkage (%)
1050	2.0	0.62
1100	2.0	0.91
1150	2.0	1.31
1200	0.5	1.05
1200	1	1.38
1200	2	1.63
1250	2	1.82

What is the apparent activation energy?

Solution: From Figure 15.11, it can be seen that a shrinkage of 1.31% will occur at $\ln(\text{time}) = -0.16$ or $t = 0.852$ hrs.

Substituting the time and temperatures for a shrinkage of 1.3%, $Q = -R\ln(t_2/t_1)/(1/T_2-1/T_1)$, $Q = -8.23\ln(2/0.852)/(1/1473-1/1423) = 294$ kJ/mole.

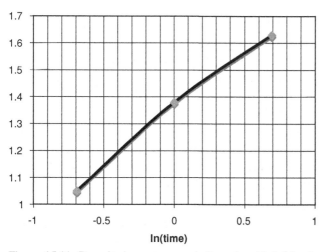

Figure 15.11. Porosity increases nearly linearly with ln(time).

Liquid Phase Sintering

Mixtures of powders of two materials sinter very rapidly if one of them melts at the sintering temperature. Initially capillary action causes the liquid phase to rapidly wet the solid phase causing an initial contraction. As the solid phase dissolves in the liquid, it is rapidly transported to locations that decrease the pore volume. Carbide tools are made from a mixture of cobalt and tungsten carbide powders, which are sintered above the melting point of cobalt. The volume fraction of liquid must be limited, so capillarity can retain the shape during sintering. Figure 15.12 shows the microstructure of a tungsten carbide tool containing 85% WC and 15% Co.

Figure 15.12. Microstructure of an 85% WC, 15% Co tool material (1500×). From American Society for Metals, *Metals Handbook*, 8th ed., v. 7, 1972.

Hot Isostatic Pressing

If pressure is applied at the sintering temperature, plastic deformation by creep increases the rate of sintering. This is referred to as hot isostatic pressing, or *HIPping*. It is used on castings as well as powder compacts.

Note of Interest

The gradual hardening of loose snow into dense snow is an example of sintering. The energy associated with surfaces is reduced as the arms of dendritic crystals become rounded and bind to other crystals. Making snowmen and snow balls can be regarded as powder processing.

Problems

1. The density of a powder after compaction is 85% and after sintering 99%. What diameter die and punch should be used to make a cylinder 25 mm in diameter and 22 mm tall?
2. The time for a given degree of sintering is proportional to D^m. Determine the exponent, m, for ice from the data in Figure 15.8.

3. Data for the neck growth of PMMA during sintering is plotted in Figure 15.13.

   A. Determine the activation energy.

   B. Determine the value of the exponent, n, in the equation $X/D = Ct^n$.

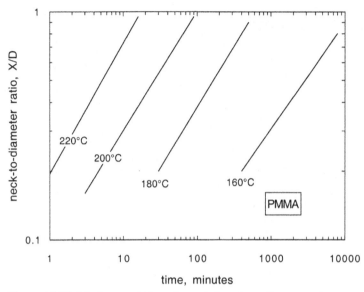

Figure 15.13. Neck growth in polymethylmethacrylate at various temperatures. Data from N. Rosenzweig and M. Narkis, *Polymer Sci. Eng.*, v. 11, 1981. From W. F. Hosford, *Materials Science: An Intermediate Text*, Cambridge, 2006.

Pottery and Concrete

Pottery was the first man-made material. The oldest known pottery is dated at 27,000 to 23,000 B.C. Pottery is made by shaping clay, drying it, and then firing it at an elevated temperature, which changes its chemical structure.

Clay

There are several forms of clay. All are aluminosilicates. Many contain other elements. Figure 16.1 shows the plate-like structure of kaolinite, $Al_2(Si_2O_5)(OH)_4$. Water absorbed between platelets allows them to slide easily over one another.

Typical clays used for ceramics often contain other materials including ground quartz (SiO_2) and a flux such as a feldspar, which is an aluminosilicate containing Na^+, K^+, and Ca^{2+}.

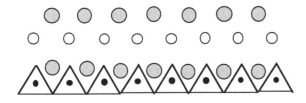

Figure 16.1. Schematic of the molecular structure of clay.

 Al^{3+}

 OH^-

 silica tetrahedron, $(Si_2O_5)^{2-}$

Processing of Clay Products

Clay is shaped either by pressing it into the desired shape or by *slip casting*. *Slip* is a suspension of clay in water made by adding a small amount of a *deflocculant* (often sodium silicate or soda ash), which allows the clay to be suspended in water. The slip is poured into a plaster of paris mold that absorbs water and causes the clay near the surface of the plaster to become solid. When this solid reaches the desired thickness, the mold is emptied and the resulting shape removed. Figure 16.2 illustrates the process.

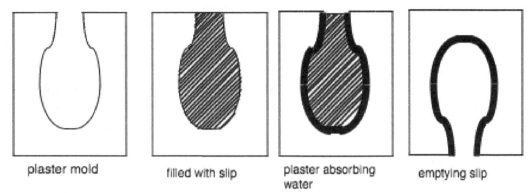

plaster mold   filled with slip   plaster absorbing water   emptying slip

Figure 16.2. Steps in slip casting.

The shape is fired only after it is allowed to air dry. Some shrinkage occurs during air drying. After air drying, the shape is said to be *green* and must be handled with care because it has little strength. Next it is fired usually between 900 and 1400 °C. Firing drives off the chemically combined water, with additional shrinkage but a great increase of strength. For pure kaolinite, the composition after firing is $3(Al_2O_3) \cdot 2(SiO_2)_2$, which is a mineral called mullite:

$$3Al_2(Si_2O_5)(OH)_4 \rightarrow 3(Al_2O_3) \cdot 2(SiO_2)_2 + 6H_2O.$$

Figure 16.3 is the SiO_2-Al_2O_3 phase diagram.

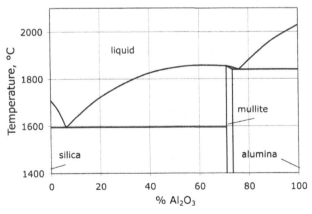

Figure 16.3. The $SiO_2 \cdot Al_2O_3$ phase diagram showing mullite as an intermediate phase.

Example Problem 16–1:

Calculate the percent weight loss when kaolinite is fired.

Solution: In the reaction $3Al_2(Si_2O_5)(OH)_4 \rightarrow 3(Al_2O_3) \cdot 2(SiO_2)_2 + 6H_2O$, the molecular weight of the clay is $6(27) + 6(14) + 27(16) + 12(1) = 690$, and the molecular weight of the water given off is $6(18) = 108$, so the loss in weight is $108/690 = 15.7\%$.

Silica and feldspar are added to the kaolinite to form a glass that bonds the mullite crystals together.

Most pottery items are glazed. After firing, powdered glass is painted onto the item's surface. On re-firing, the glaze melts and forms a continuous layer over the surface.

Forms of Pottery

Earthenware, ovenware, stoneware, china, and porcelain are various forms of pottery that differ in composition and firing temperature. Earthenware is basically kaolinite with some silica and feldspar fired at lower temperatures than other clay products. It is relatively weak with a porosity of 5 to 20%. It may be left unglazed, as in flowerpots and soil pipe, or it may be glazed to make it water resistant.

Stoneware is fired at higher temperatures. This reduces the porosity to less than 5% giving it better strength. China is fired at even higher temperatures, which converts much of the silica, clay, and feldspar mixture into a glass. Increased firing temperatures lower the thermal expansion coefficient, α. Pottery with $\alpha < 4 \times 10^{-6}$ /K can withstand air cooling after heating in an oven. To be able to withstand flame contact, the coefficient of thermal expansion should be less than 2×10^{-6} /K.

Most pottery is glazed. A glaze is applied after firing as a suspension of low-melting glass particles and then the pottery is fired again to allow the glaze to melt and spread over the surface. Glazing has two functions. Earthenware vessels would otherwise be unable to hold liquids due to porosity. Glazing is also done for decoration. To avoid cracking on cooling, the glaze should have a lower coefficient of thermal expansion than the pottery base so that it will be left under a residual compressive stress. Sometimes, for aesthetic reasons, glazes are intentionally formulated to produce surface cracks.

Hydration Reactions

Several materials react with water to form solids. Among these are plaster of paris, lime plaster, and portland cement. Plaster of paris is the hemi-hydrate of calcium sulfate, $CaSO_4 \cdot 0.5H_2O$. It is obtained by heating gypsum, $CaSO_4 \cdot 4H_2O$, to about $150\,°C$, which releases some of the hydrated water:

$$2CaSO_4 \cdot 4H_2O \rightarrow 2CaSO_4 \cdot H_2O + 3H_2O.$$

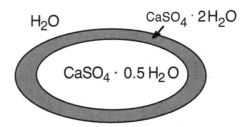

Figure 16.4. In a typical hydration reaction, the product separates the reacting chemicals. Here water is separated from the plaster of paris by the gypsum formed.

When dry plaster powder is mixed with water to form a paste, it hydrates to form hard gypsum by the reverse reaction. The rate of the hydration reaction depends on the particle size of the plaster of paris because the reaction product surrounds the unreacted plaster of paris as shown in Figure 16.4.

Lime plaster is a mixture of calcium hydroxide and sand (or other inert fillers). Lime is made by heating limestone (calcium carbonate) to above 825 °C to cause it to decompose into calcium oxide and carbon dioxide:

$$CaCO_3 \rightarrow CaO + CO_2.$$

When CaO is hydrated with water it forms *slaked lime* or lime plaster:

$$CaO + H_2O \rightarrow Ca(OH)_2.$$

With long periods of time, plaster reacts with atmospheric CO_2 to form calcium carbonate. Commercial plasters contain both plaster of paris and lime. Lime is often added to plaster of paris to control the rate of hardening. The rate of the hydration reaction depends on the particle size as well as on the fraction of each component.

Portland Cement

Portland cement contains several minerals. The two principal minerals are tricalcium silicate, $(CaO)_3 \cdot (SiO_2)$, which is abbreviated by C_3S, and dicalcium silicate, $(CaO)_2 \cdot (SiO_2)$, abbreviated by C_2S. There is also some tricalcium aluminate, $(CaO)_3 \cdot (Al_2O_3)$, abbreviated by C_3A.

Dry powdered cement is mixed with water, which forms hydrates by the reactions

$$Ca_3Al_2O_6 + 6H_2O \rightarrow Ca_3Al_2(OH)_{12},$$
$$Ca_2SiO_4 + xH_2O \rightarrow Ca_2SiO_4 \cdot (H_2O)_x,$$
$$Ca_3SiO_5 + (x+1)H_2O \rightarrow Ca_2SiO_4 \cdot (H_2O)_x + Ca(OH)_2.$$

Cement may also contain some tricalcium aluminum ferrite, $(CaO)_3 \cdot (Al_2O_3) \cdot (FeO)$, abbreviated by C_3AF.

Hardening occurs by these hydration reactions in which water is chemically bound into hard compounds rather than by "drying." The rate of hardening is controlled by the composition of the mixture and by the particle size of the cement powder. Because heat is released during hydration, large structures must be cooled to slow the curing and prevent thermal gradients that would cause cracking.

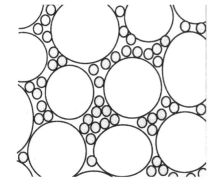

Figure 16.5. Concrete consists of gravel particles (large open circles), sand (small gray circles), and cement paste filling the remaining space.

Concrete Mixes

Concrete is a mixture of gravel, sand, and cement paste (Figure 16.5). There are two reasons for adding gravel and sand to cement. Sand and gravel particles are stronger than the hydrated cement, and they are less expensive than cement. If the particles of gravel filled 60% of the space and sand filled 60% of the remaining space, the volume of the cement paste would be only (100–60)(100–40) = 16%. However, because mixing is not complete, the ratio of sand to cement and the ratio of cement paste to sand should be higher.

A common mix of gravel, sand, cement, and water in the ratio of 3.1:2.6:1.0:0.55, respectively, by volume has enough sand to fill the spaces between gravel particles and enough cement paste to fill the spaces between the sand particles. Higher ratios of sand to gravel and cement to sand will be more expensive.

The ratio of water to cement determines the final strength. Excess water that does not enter the hydration reactions must eventually evaporate. In doing so it may leave hairline cracks. However, the mix must contain enough water to make the concrete workable. Figure 16.6 shows how the strength of concrete varies with the amount of water.

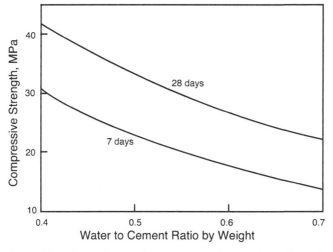

Figure 16.6. Dependence of the strength of concrete on the ratio of water to cement.

Example Problem 16–2:

If 1 m^3 of gravel weighs 1.8 Mg and the density of the gravel particles is 2.8, what is the volume of open space?

Solution: The open volume is $1 - 1.6/2.8 = 0.429$.

Asphalt Concrete

Asphalt concrete, commonly called asphalt, consists of asphalt binder mixed with an aggregate (usually sand). Mixing of asphalt and aggregate is accomplished in one of several ways. The asphalt may be heated to about 150 °C to decrease its viscosity, or its viscosity may be lowered by mixing with waxes or by emulsifying with water or other materials. Asphalt concrete is used as a paving material.

Notes of Interest

The earliest pottery was fired in bonfires for short times. The highest temperatures were probably about 600 °C. Sand, grit, crushed shell, and crushed pottery were often added to the clay. These particles in the clay decreased the shrinkage during cooling, and reduced thermal stress and cracking. Bonfire-fired pots had round bottoms, which reduced their tendency to crack. Holes dug in the ground and covered with fuel were the first kilns, which allowed better control over firing. The potter's wheel was invented in Mesopotamia sometime between 6000 and 4000 B.C. It permitted potters to meet the expanding needs of the first cities. Pottery was in use in India as early as 5000 B.C. in the Indus Valley. The Greeks decorated amphoras with geometric designs as early as 1000 B.C.

In the construction of the Hoover Dam, water pipes were incorporated into the concrete to remove the heat produced by the hydration reactions.

Problems

1. The composition $(Al_2O_3) \cdot 2(SiO_2)$ that results from the firing of kaolinite contains what weight percent SiO_2?
2. If the fraction of space filled by one cubic meter of gravel is 58%, the fraction of space filled by a batch of sand is 48%, and just enough sand were added to the gravel to fill the space between the gravel particles, what would be the volume of the pores left?
3. What should be the relation between the thermal expansion coefficients of a glaze and the pottery to ensure that a glaze is not cracked?

17 Composites

Composite materials have been used throughout history to achieve combinations of properties that could not be achieved with individual materials. Concrete is a composite of cement, sand, and gravel. Poured concrete is usually reinforced with steel rods. Other examples of composites include steel-belted tires; asphalt blended with gravel for roads; plywood with alternating directions of fibers, carbon, or glass fiber-reinforced polyester; or epoxy used for furniture, boats, and sporting goods. Composite materials offer combinations of properties otherwise unavailable. The reinforcing material may be in the form of particles, fibers, or sheet laminates.

Fiber-reinforced Composites

Fiber composites may also be classified according to the nature of the matrix and the fiber. Examples of a number of possibilities are listed in Table 17.1.

Different geometric arrangements of the fibers are possible. In two-dimensional products, the fibers may be unidirectionally aligned at 90° to one another in a woven fabric or randomly oriented (Figure 17.1). The fibers may be very long or chopped into short segments for easy fabrication. In thick sections it is possible to randomly orient short fibers in three dimensions. Fiber reinforcement is used to impart stiffness (increased modulus) or strength to the matrix. Fiber reinforcement also increases toughness.

Table 17.1. *Examples of fiber-reinforced composites*

Fiber	Metal matrix	Ceramic matrix	Polymer matrix
Metal	Al/W	concrete/steel	rubber(tires)/steel
Ceramic	Al/SiC	Sialon/SiC	polyester/fiber glass
Polymer	–	clay/straw	epoxy/Kevlar

Elastic Behavior of Fiber-reinforced Composites

The strains parallel to long parallel fibers must be the same in both the matrix and the fiber: $\varepsilon_f = \varepsilon_m = \varepsilon$. For loading parallel to the fibers, the total load, F, is the sum

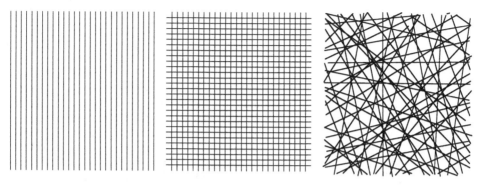

Figure 17.1. Several geometric arrangements of fiber reinforcements. From W. F. Hosford, *Mechanical Behavior of Materials*, Cambridge, 2005.

of the forces on the fibers, F_f, and the matrix, F_m. The forces expressed in terms of the stresses are $F_f = \sigma_f A_f$ and $F_m = \sigma_m A_m$, where σ_f and σ_m are the stresses in the fiber and matrix and where A_f and A_m are the cross-sectional areas of the fiber and matrix. The total force, F, is the sum of $F_f + F_m$ so

$$\sigma A = \sigma_f A_f + \sigma_m A_m,\qquad\qquad 17.1$$

where A is the overall area, $A = A_f + A_m$, and σ is the overall stress, F/A. For elastic loading, $\sigma = E\varepsilon$, $\sigma_f = E_f \varepsilon_f$, and $\sigma_m = E_m \varepsilon_m$, so $E\varepsilon A = E_f \varepsilon_f A_f + E_m \varepsilon_m A_m$. Realizing that $\varepsilon_f = \varepsilon_m = \varepsilon$, and expressing $A_f/A = V_f$ and $A_m/A = V_m$,

$$E = E_f V_f + E_m V_m.\qquad\qquad 17.2$$

This is often called the *rule of mixtures*. It is an upper bound to the elastic modulus of a composite.

Example Problem 17–1:

Calculate the Young's modulus of a one-dimensional fiber-reinforced composite with 35% volume glass fibers in a polyester matrix.

Solution: Using equation 17.2, $E = E_f V_f + E_m V_m$, with $E_f = 70$ MPa, $E_m = 3$ MPa, $V_f = 0.35$, and $V_m = 0.65$, $E = 0.35(70) + 0.65(3) = 16$ MPa.

Consider the behavior of the same composite under tension perpendicular to the fibers. Now one cannot assume that $\varepsilon_f = \varepsilon_m = \varepsilon$. An alternative, albeit extreme, assumption is that the stresses in the matrix and the fibers are the same: $\sigma_f = \sigma_m = \sigma$. In this case $\varepsilon = \sigma/E$, $\varepsilon_f = \sigma_f/E_f$, and $\varepsilon_m = \sigma_m E_m$ and the overall (average) strain is $\varepsilon = \varepsilon_f V_f + \varepsilon_m V_m$. Combining, $\sigma/E = V_f \sigma_f/E_f + V_m \sigma_m E_m$. Finally realizing that $\sigma_f = \sigma_m = \sigma$, we have

$$1/E = V_f/E_f + V_m/E_m.\qquad\qquad 17.3$$

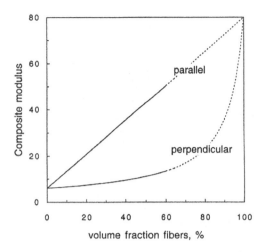

Figure 17.2. Upper and lower bounds to Young's modulus for composites. The upper bound is appropriate for loading parallel to the fibers. Loading perpendicular to the fibers is between these two extremes. The lines are dashed above $V_f = 60\%$ because this is a practical upper limit to the volume fraction of fibers. From W. F. Hosford, *Mechanical Behavior of Materials*, Cambridge, 2005.

Equation 17.3 is a lower bound for the modulus. Figure 17.2 shows the predictions of equations 17.2 and 17.3. The actual behavior for loading perpendicular to the fibers is between these two extremes.

Now consider the orientation dependence of the elastic modulus of a composite with unidirectionally aligned fibers. Let 1 be the axis parallel to the fibers with axes 2 and 3 perpendicular to the fibers. For a uniaxial stress applied along a direction x at an angle θ from axis 1 and $90 - \theta$ from axis 2, the stresses on the three-axis system may be expressed as

$$\sigma_1 = \cos^2\theta\sigma_x,$$
$$\sigma_2 = \sigma_3 = \sin^2\theta\sigma_x,$$
$$\tau_{12} = \sin\theta\cos\theta\sigma_x, \qquad\qquad 17.4$$
$$\sigma_3 = \tau_{23} = \tau_{31} = 0.$$

Hooke's laws give the strains along axes 1 and 2:

$$e_1 = (1/E_1)[\sigma_1 - \upsilon_{12}\sigma_2], e_2 = (1/E_2)[\sigma_2 - \upsilon_{12}\sigma_1] \quad \text{and}$$
$$\gamma_{12} = (\tau_{12}/G_{12}). \qquad\qquad 17.5$$

The strain in the x direction can be written as

$$e_x = e_1\cos^2\theta + e_2\sin^2\theta + 2\gamma_{12}\cos\theta\sin\theta$$
$$= (\sigma_x/E_1)[\cos^4\theta - \upsilon_{12}\cos^2\theta\sin^2\theta] + (\sigma_x/E_2)[\sin^4\theta - \upsilon_{12}\cos^2\theta\sin^2\theta] \quad 17.6$$
$$+ (2\sigma_x/G_{12})\cos^2\theta\sin^2\theta.$$

Figure 17.3 shows that the modulus drops rapidly for off-axis loading. The average for all orientations in this figure is about 18% of E_1.

A crude estimate of the effect of a cross-ply can be made from an average of the stiffnesses due to fibers at $0°$ and $90°$ as shown in Figure 17.4. While the cross-ply stiffens the composite for loading near $90°$, it has no effect on the $45°$ stiffness. Figure 17.5 illustrates why this is so. For loading at $45°$, small extensions can be accommodated by rotation of the fibers without any extension.

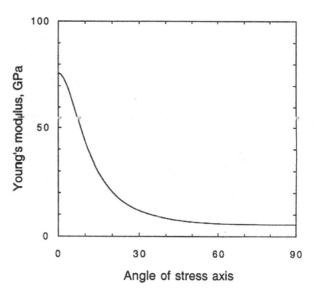

Figure 17.3. The orientation dependence of the elastic modulus in a composite with unidirectionally aligned fibers. Here it is assumed that $E_1 = 75$ GPa, $E_2 = 5$ GPa, $G = 10$ GPa, and $\upsilon_{12} = 0.3$. Note that most of the stiffening is lost if the loading axis is misoriented from the fiber axis by as little as 15°. From W. F. Hosford, *Mechanical Behavior of Materials*, Cambridge, 2005.

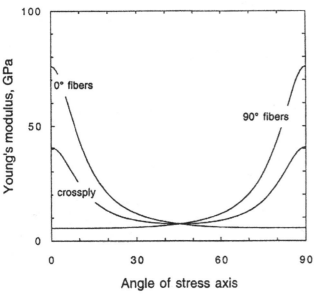

Figure 17.4. Calculated orientation dependence of Young's modulus in composites with singly and biaxially oriented fibers. For this calculation, it was assumed that $E_\perp = 75$ GPa and $E_{//} = 5$ GPa. Note that even with biaxially oriented fibers, the modulus at 45° is very low. From W. F. Hosford, *ibid*.

The simple averaging used to calculate the curves in Figure 17.4 underestimates the stiffness for most angles of loading, θ. It assumes equal strains in both plies in the loading direction but neglects the fact that the strains perpendicular to the loading direction in both plies must also be equal. When this constraint is accounted for, a somewhat higher modulus is predicted. This effect, however, disappears at $\theta = 45°$ because composites with both sets of fibers have the same lateral strain.

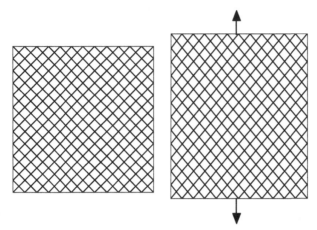

Figure 17.5. Rotation of fibers in a woven cloth. Stress at 45° to the fibers allows deformation with little or no stretching of the fibers.

With randomly oriented fibers, the orientation dependence disappears. One might expect the modulus would be the average of the moduli for all directions of uniaxially aligned fibers. However, this again would be an underestimation since it neglects the fact that the lateral strains must be the same for all fiber alignments. A useful engineering approximation for randomly aligned fibers is

$$E \approx (3/8)E_\perp + (5/8)E_\parallel, \qquad\qquad 17.7$$

where $E_\perp$ and $E_\parallel$ are the moduli perpendicular and parallel to uniaxially aligned fibers.

Strength of Fiber-reinforced Composites

The rule of mixtures cannot be used to predict the strengths of composites with uniaxially aligned fibers. The reason can be appreciated by considering the stress-strain behavior of both materials as shown schematically in Figure 17.6. The strains in the matrix and fibers are equal so the fibers reach their breaking strengths long before the matrix reaches its tensile strength. Thus, the strength of the composite is $UTS < V_m UTS_m + V_f UTS_f$.

Usually the load carried by the fibers is greater than the breaking load of the matrix, so it will fail when the fibers break. The composite strength is given by

$$UTS = V_m \sigma_m + V_f (UTS)_f, \qquad\qquad 17.8$$

where $\sigma_m = (E_m/E_f)(UTS)_f$ is the stress carried by the matrix when the fiber fractures.

For composites with low volume fraction fibers, the fibers may break at a load less than the failure load of the matrix. In this case, after the fibers break, the whole load must be carried by the matrix so the predicted strength is

$$UTS = V_m (UTS)_m. \qquad\qquad 17.9$$

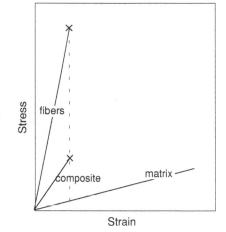

Figure 17.6. Stress-strain curves for the matrix, the fibers, and the composite.

Volume Fraction Fibers

The stiffness and strength of reinforced composites should increase with volume fraction of fibers, but there are practical limitations on the volume fraction. Fibers must be separated from one another. Fibers are often pre-coated to ensure this separation and to control the bonding between fibers and matrix. Techniques of infiltrating fiber arrangements with liquid resins lead to variability in fiber spacing as shown in Figure 17.7. The maximum possible packing density is greater for unidirectionally aligned fibers than for woven or cross-ply reinforcement. A practical upper limit for volume fraction seems to be about 55 to 60%. This is why the calculated lines in Figures 17.2 and 17.6 are dashed above $V_f = 60\%$.

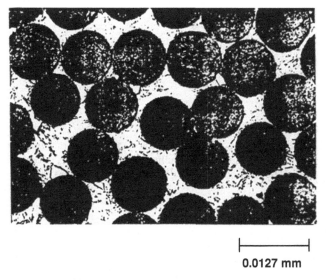

0.0127 mm

Figure 17.7. Glass fibers in a polyester matrix. Note the variability in fiber spacing. From *Engineered Materials Handbook, v. 1, Composites*, ASM International, 1987.

Fiber Length

Fabrication is much simplified if the reinforcement is in the form of chopped fibers. Chopped fibers can be blown onto a surface to form a mat. Composites with chopped fibers can be fabricated by processes that are impossible with continuous fibers, such as extrusion, injection molding, and transfer molding. The disadvantage of chopped fibers is that some of the reinforcing effect of the fibers is sacrificed because the average axial stress carried by fibers is less for short ones than long ones. The reason is that, at the end of the fiber, the stress carried by the fiber is vanishingly low. Stress is transferred from the matrix to the fibers primarily by shear stresses at their interfaces. The average axial stress in a fiber depends on its aspect ratio, D/L, where D and L are the fiber's diameter and length.

Failures with Discontinuous Fibers

Failure may occur either by fracture of fibers or by the fibers pulling out of the matrix. Both possibilities are shown in Figure 17.8. Pullout will occur if the plane of the crack is near the end of the fiber. If it is not near the end, the fiber will fracture. Figure 17.9 is a picture showing the pullout of boron fibers in an aluminum matrix. To fracture a fiber of diameter D, the force, F, must be

$$F = \sigma^* \pi D^2/4, \qquad\qquad 17.10$$

where σ^* is the fracture strength of the fiber.

Much more energy is absorbed if the fibers pull out. Therefore greater toughness is often achieved with shorter fibers and lower fiber-matrix interface strength. This is illustrated in Figure 17.10.

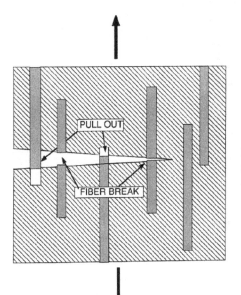

Figure 17.8. Sketch showing some fibers fracturing at a crack and others pulling out. From W. F. Hosford, *Mechanical Behavior of Materials*, Cambridge, 2005.

Figure 17.9. Photograph of SiC fibers pulling out of a titanium matrix. From T. W. Clyne and P. J. Withers, *An Introduction to Metal Matrix Composites*, Cambridge, 1993.

Typical Properties

Two types of polymer matrices are common: epoxy and polyester. Most of the polymers used for matrix materials have moduli of 2 to 3 GPa and tensile strengths in the range of 35 to 70 MPa. Fiber reinforcements include glass, boron, Kevlar, and carbon. Properties of some epoxy matrix composite systems are given in Table 17.2. Properties of some commonly used fibers are given in Table 17.3.

Other fiber composites include ceramics reinforced with metal or ceramic fibers. Metals such as aluminum-base alloys may be reinforced with ceramic fibers to increase their stiffness. In some eutectic systems, directional solidification can lead to rods of one phase reinforcing the matrix.

Particulate Composites

Composites reinforced by particles rather than long fibers include diverse materials such as concrete (cement matrix with sand and gravel particles), polymers filled with wood flour, and carbide tools with a cobalt-base matrix alloy hardened by tungsten carbide particles. Sometimes the purpose is simply economics (e.g., wood flour is cheaper than plastics). Another objective may be increased hardness (e.g., carbide

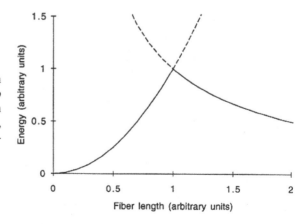

Figure 17.10. The energy for expended in fiber pullout increases with fiber length up to a critical length and then decreases with further length increase. From W. F. Hosford, *Mechanical Behavior of Materials*, Cambridge, 2005.

Table 17.2. *Properties of epoxy matrix composites*

Fiber	Vol % fiber	Young's modulus (GPa)		Tensile strength (MPa)	
		Transverse	Longitudinal	Transverse	Longitudinal
E glass unidirectional	60	40	10	780	28
E glass bidirectional	35	16.5	16.5	280	280
E glass chopped matte	20	7	7	100	100
Boron unidirectional	60	215	24	1400	65
Kevlar29 unidirectional	60	50	5	1350	–
Kevlar49 unidirectional	60	76	6	1350	30
Carbon	62	145		1850	

tools). The isostrain and isostress models (equations 17.2 and 17.3) are upper and lower bounds for the dependence of Young's modulus on volume fraction. The behavior of particulate composites is intermediate and can be represented by a generalized rule of mixtures of the form

$$E^n = V_A E_A^n + V_B E_B^n, \qquad 17.11$$

where A and B refer to the two phases. The exponent n lies between the extremes of $n = +1$ for the isostrain model and $n = -1$ for the isostress model. Figure 17.11 shows the dependence of E on volume fraction for several values of n. If the modulus of the continuous phase is much higher than that of the particles, $n = 1/2$ is a reasonable approximation. For high-modulus particles in a low-modulus matrix, $n < 0$ is a better approximation.

Table 17.3. *Typical fiber properties*

Fiber	Young's modulus (GPa)	Tensile strength (GPa)	Elongation (%)
Carbon (PAN* HS)	250.	2.7	1
Carbon (PAN* HM)	390	2.2	0.5
SiC	70		
Steel	210	2.5	
E-glass	70	1.75	
B	390	2 to 6	
Kevlar 29	65	2.8	4
Kevlar 49	125	2.8	2.3
Al_2O_3	379	1.4	
β-SiC	430	3.5	

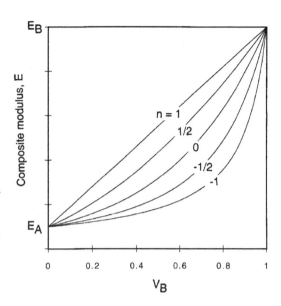

Figure 17.11. Dependence of Young's modulus on volume fraction according to equation 17.11 for several values of n. The subscripts A and B represent the lower and higher modulus phases. The isostrain and isostress models correspond to $n = +1$ and -1. Here the ratio of $E_B/E_A = 10$. From W. F. Hosford, *Mechanical Behavior of Materials*, Cambridge, 2005.

Example Problem 17–2:

Estimate Young's modulus for aluminum reinforced with 40 volume % silicon carbide. The elastic moduli are 440 MPa for SiC and 70 MPa for aluminum. Assume that $n = 0.5$.

Solution: Using equation 15.11, $E^n = V_A E_A^n + V_B E_B^n$, with $n = 0.5$, $E = [(0.4)(430)^{0.5} + (0.60)(70)^{0.5}]^2 = 177$ MPa.

Lamellar Composites

Two or more sheets of materials bonded together can be considered lamellar composites.

Examples include safety glass, plywood, plated metals, and glazed ceramics. Consider sheets of two materials, A and B, bonded together in the x-y plane and loaded in this plane. The basic equations governing the strain are

$$\varepsilon_{xA} = \varepsilon_{xB} \text{ and } \varepsilon_{yA} = \varepsilon_{yB}. \qquad 17.12$$

If both materials are isotropic and the loading is elastic, equations 17.12 become

$$\varepsilon_{xA} = (1/E_A)(\sigma_{xA} - \upsilon_A \sigma_{yA}) = \varepsilon_{xB} = (1/E_B)(\sigma_{xB} - \upsilon_B \sigma_{yB})$$
$$\varepsilon_{yA} = (1/E_A)(\sigma_{yA} - \upsilon_A \sigma_{xA}) = \varepsilon_{yB} = (1/E_B)(\sigma_{yB} - \upsilon_A \sigma_{xB}). \qquad 17.13$$

The stresses are

$$\sigma_{xA} t_A + \sigma_{xB} - t_B = \sigma_{xav},$$
$$\sigma_{yA} t_A + \sigma_{yB} t_B = \sigma_{yav}, \qquad 17.14$$
$$\sigma_{zA} = \sigma_z = 0,$$

where t_A and t_B are the fractional thicknesses of A and B.

Now consider loading under uniaxial tension applied in the x direction. Substituting $\sigma_{yav} = 0, \sigma_{yB} = -(t_A/t_B)\sigma_x$ and $\sigma_{xB} = (\sigma_{xav} - t_A\sigma_{xA})/t_B$ into equation 17.14,

$$\sigma_{xA} - \upsilon_A\sigma_{yA} = (E_A/E_B)[(\sigma_{xav} - t_A\sigma_{xA})/t_B + \upsilon_B(t_A/t_B)\sigma_{yA}] \quad \text{and}$$

$$\sigma_{yA} - \upsilon_A\sigma_{xA} = (E_A/E_B)[(V_A/V_B)\sigma_{yA} + \upsilon_B(\sigma_{xav} - t_A\sigma_{xA})/t_B]. \qquad 17.15$$

Young's modulus according to an upper-bound isostrain model for loading in the plane of the sheet can be expressed as

$$E = N/D \text{ where}$$
$$N = E_B^2 t_B^2(1 - \upsilon_A^2) + 2E_A E_B t_A t_B(1 - \upsilon_A\upsilon_B) + E_A^2 t_A^2(1 - \upsilon_B^2) \quad \text{and}$$
$$D = E_B t_B(1 - \upsilon_A^2) + E_A t_A(1 - \upsilon_B^2). \qquad 17.16$$

For the special case in which $\upsilon_B = \upsilon_A$, these expressions reduce to the upper-bound model

$$Et = E_B t_B + E_A t_A. \qquad 17.17$$

Notes of Interest

The use of reinforced concrete is usually attributed to Joseph-Louis Lambot in 1448. In 1868 Joseph Monier, a French gardener, patented a design for reinforced garden tubs and later patented reinforced concrete beams and posts for railway and road guardrails.

Fiberglass-reinforced plastics were developed during World War II for use in radar domes. The first main use for commercial products was in the 1950s.

Problems

1. A 1-mm-diameter rod of glass A is coated with 1 μm of glass B. The properties of the glasses are

Glass	Coefficient of thermal expansion ($°C^{-1}$)	Young's modulus (GPa)
A	9×10^{-6}	70
B	1×10^{-6}	70

If there are no stresses in the rod and coating at 200 °C, what are the stresses in the rod and coating when they are cooled to 20 °C? Indicate whether they are tensile or compressive.

2. Consider a carbon-reinforced epoxy composite containing 45 volume % unidirectionally aligned carbon fibers. Calculate the composite modulus and then calculate the composite tensile strength. Assume both epoxy and carbon are elastic to fracture.

	Young's modulus	Tensile strength
Epoxy	3 GPa	55 MPa
Carbon	250 GPa	2.5 GPa

3. Carbide cutting tools are composites of very hard tungsten carbide particles in a cobalt matrix. The elastic moduli of tungsten carbide and cobalt are 102×10^6 and 30×10^6 psi, respectively. It was experimentally found that the elastic modulus of a composite containing 52 volume % carbide was 60×10^6 psi. What value of the exponent, $E_A t_A$, in equation 17.11 would this measurement suggest? A trial-and-error solution is necessary to solve this. (Note that $E_A t_A = 0$ is a trivial solution.)

4. In all useful fiber-reinforced composites, the elastic moduli of the fibers are higher than those of the matrix. Explain why.

5. Calculate the maximum possible volume fraction fibers for:
   a. Unidirectionally aligned cylindrical fibers.
   b. Unidirectionally aligned cylindrical fibers of diameter 100 μm coated with a 5 μm layer.

18 Carbon

Carbon can occur in several different forms including diamond, graphite, amorphous carbon, and fullerenes. None of these forms fit into the classification of materials as metals, ceramics, or polymers. Figure 18.1 shows the equilibrium between graphite, diamond, and liquid.

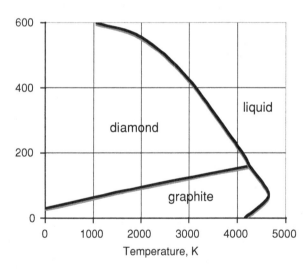

Figure 18.1. Phase diagram showing the equilibrium phases of carbon. Data from C. G. Suits, *American Scientist*, v. 52, p. 395, 1964.

Diamond

Each carbon atom in diamond is covalently bonded to four other carbon atoms as shown in Figure 18.2. Very strong bonding makes diamond the hardest material known (10,000 kg/mm^2). Diamond is used for cutting very hard materials. Diamond has an extremely high Young's modulus (1,050 GPa) and a very low coefficient of thermal expansion (1 × 10^{-6}/K). It has the highest thermal conductivity of all materials (2 kW/m-K compared with 401W/m-K for copper), making it useful for dissipating heat. Its density $(3.52\,\mathrm{Mg/m^3})$ is considerably greater than that of graphite $(2.25\,\mathrm{Mg/m^3})$.

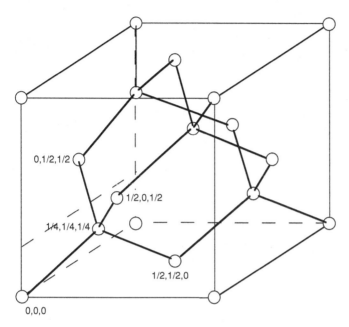

Figure 18.2. The crystal structure of diamond. Each carbon atom is covalently bonded to four others. From W. F. Hosford, *Materials Science: An Intermediate Text*, Cambridge, 2006.

The first synthetic diamonds were made by subjecting carbon to very high pressures at high temperatures. Diamond can also be grown by chemical vapor deposition (CVD) under low pressure (1 to 27 kPa). Gasses include a carbon source and typically hydrogen heated in a pressurized chamber and broken down, depositing diamond on exposed surfaces. Large areas ($> 150\,\mathrm{mm^2}$) can be coated on a substrate. This allows CVD diamond films to be used as heat sinks in electronics and to be used in wear-resistant surfaces.

Graphite

The structure of graphite consists of sheets of carbon atoms arranged in a hexagonal pattern (Figure 18.3). The bonding in the hexagonal sheets is like that in a benzene ring. Each carbon has two single bonds and one double bond. The sheets are bound to each other by weak van der Waals bonds. The ease with which sheets can slide over one another explains the lubricity of graphite. Because the double bond can move freely, the electrical conductivity in the plane of the sheet is very high, like that of a metal. The electrical and thermal conductivities perpendicular to the sheets are very low. The Young's modulus is very high in the planes of the sheet and very low perpendicular to them. The anisotropy of properties listed in Table 18.1 reflects the difference in bond strengths parallel and perpendicular to the sheets.

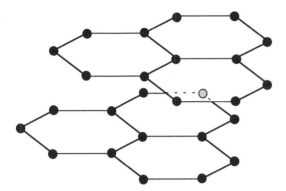

Figure 18.3. The structure of graphite. From W. F. Hosford, *Materials Science: An Intermediate Text*, Cambridge, 2006.

Table 18.1. *Directional properties of graphite**

Property	Perpendicular to c	Parallel to c
Electrical resistivity (ohm-m)	2.5 to 5 $\times$ 10^{-6}	3000 $\times$ 10^{-6}
Thermal conductivity (W/m-K)	398	2.2
Thermal expansion at 20°C ($°C^{-1}$)	25 $\times$ 10^{-6}	slightly negative ($°C^{-1}$)
Elastic modulus (GPa)	1060	35.5

* Data from H. O. Pierson, *Handbook of Carbon, Graphite, Diamond and Fullerenes*, Knovel, 2001.

Carbon Fibers

Carbon fibers are thin graphite ribbons. They are made by pyrolizing polymeric precursors such as polyacrylonitrile, pitch, rayon, or other polymers that have carbon-carbon backbones. Processing consist of several steps. The first is stretching to align molecular chains; then they are heated to stabilize the orientation, pyrolize, and graphitize. The strengths and moduli are very high because they involve stretching carbon-carbon bonds. The properties of graphite fibers depend greatly on the nature of the precursor, its diameter, and the details of processing. Young's moduli of commercial carbon fibers vary from 200 to 700 GPa and tensile strengths from 2 to 7 GPa. Carbon fibers are used in epoxy-bonded composites.

Amorphous Carbon

Although soot and coal are often referred to as amorphous, they actually contain small regions that are graphitic or diamond-like. True amorphous carbon can be produced by vapor sputter deposition. Some of the bonding is sp^2 (graphite-like) and some is sp^3 (diamond-like). The ratio of the two types of bonds may vary considerably. Amorphous carbon contains a high concentration of dangling bonds. These cause deviations of the interatomic spacing of more than 5% and noticeable variations in bond angles.

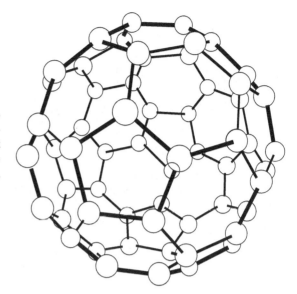

Figure 18.4. C_{60} buckyball. There are 60 carbon atoms arranged in hexagons and pentagons. The arrangement is the same as that on a soccer ball. From W. F. Hosford, *Materials Science: An Intermediate Text*, Cambridge, 2006.

Fullerenes

Until 1985, the only known elemental forms of carbon were diamond, graphite, and amorphous carbon. Then Kroto et al.* announced the discovery of C_{60}, a spherical arrangement of carbon atoms in hexagons and pentagons as shown in Figure 18.4. They called this form *Buckminsterfullerene* after the architect Buckminster Fuller, who developed the geodesic dome. The name for this type of carbon molecule has since been shortened to *fullerene*, but the molecules are commonly called *buckyballs*. Since their first discovery, it has been found that fullerenes can be made in quantity from electrical arcs between graphite electrodes. About 75% of the fullerenes produced by arcs are C_{60}, 23% C_{80}, with the rest being even larger molecules. About 9,000 fullerene compounds are known.

Nanotubes

A nanotube can be regarded as a hexagonal sheet of carbon atoms (*graphene* sheet), rolled up to make a cylinder and capped at the ends by half of a buckyball as illustrated in Figure 18.5. Tubes typically have diameters of about 1 nm. The diameter of the smallest nanotube corresponds to the diameter of the smallest buckyball (C_{60}). The length-to-diameter ratio is typically about 10^4.

Nanotubes fall into three groups, depending on the chiral angles, θ, between the sides of the hexagons and the tube axis (Figure 18.6). If $\theta = 0$, a zig-zag nanotube results. If $\theta = 30°$, the nanotube is called an armchair. Chiral nanotubes are those for which $0 < \theta < 30°$. These develop twists. Some nanotubes can have metallic conduction; others are semiconductors or insulators.

---

* H. Kroto, J. Heath, S. O'Brien, R. Curl, and R. Smalley, *Nature*, v. 318, 1985.

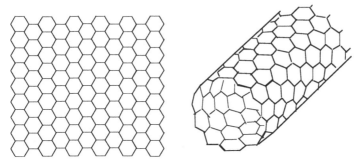

Figure 18.5. Single-wall nanotubes can be thought of being made from rolled up chicken wire. From W. F. Hosford, *Materials Science: An Intermediate Text*, Cambridge, 2006.

Nanotubes can be formed into rings with the two ends joined together. Concentric multiwall nanotubes can form as well as single-thickness nanotubes.

Nanotubes can be made by arc discharge, laser ablation, or by CVD. There are a number of potential uses of fullerenes. One potential use of nanotubes is for field effect transistors. Nanotubes decorated with metal atoms have a great potential for hydrogen storage for fuel cells. A_3C_{60} compounds, where A is an alkali (K, Rb, Cs Na), are superconductors. Sieves have been suggested that allow biological compounds to pass through but do not permit entrance of larger viruses.

Notes of Interest

In H. G. Wells' short story *The Diamond Maker*, he wrote about making diamonds from coal. However, the first commercial synthesis of diamonds wasn't until 1954 by General Electric. One method uses a pressure of 5 GPa at 1500°C to simulate the conditions that lead to the formation of natural diamonds. In the 1980s, CVD was developed. It uses a carbon-bearing plasma to deposit diamond on a substrate. It is limited to very thin films but has application for forming wear-resistant surfaces on other materials.

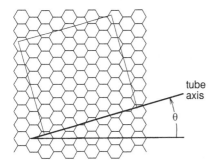

Figure 18.6. One characteristic of a nanotube is the angle between the tube axis and the crystallographic axes of the hexagons. From W. F. Hosford, *ibid*.

Problems

1. Calculate the bonding energy per mole of carbon atoms in diamond and compare this with the bonding energy per mole of carbon atoms in graphite. (Assume that in graphite each carbon atom shares two single bonds and one double bond.) Which structure, diamond or graphite, has the lower energy? Refer to Appendix 6 for bond energies.

2. Euler showed that, for a three-dimensional body, $C - E + F = 2$. A buckyball is composed of hexagons and pentagons. Three edges meet at a corner and each edge has two ends so $E = 3/2C$. Calculate the number of hexagonal and pentagonal faces in a C_{80} molecule.

19 Fibers, Foams, and Porous Materials

Fibers

Fibers of nylon, polyester, and other thermoplastics are made by extruding molten material through tiny holes in a *spinneret*. The resulting fibers are cooled before coiling. The strengths of nylon, polyester, polypropylene, and high-density polyethylene fibers are increased greatly by stretching them by 400 to 500% in tension (*drawing*) to orient the molecules parallel to the fiber axis.

Fiber strength is often quoted in terms of *grams force per denier* (g force/denier). A denier is defined as grams mass per 9000 m of fiber. Figure 19.1 shows the relative strengths of various fibers. Kevlar fibers are much stronger, having strengths of about 22 g force/denier.

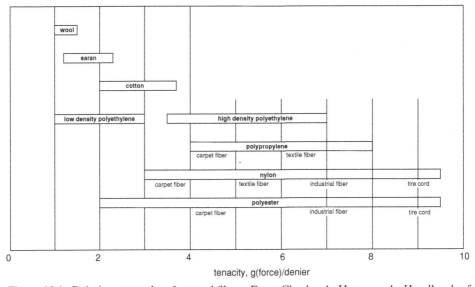

Figure 19.1. Relative strengths of several fibers. From Charles A. Harper, ed., *Handbook of Plastics and Elastomers*, McGraw-Hill, 1975.

Example Problem 19–1:

Develop an equation for converting tenacity in g force/denier to MPa.

Solution: Let T be the tenacity in g force/denier. Then T[g force/(g mass/9000 m)]ρ(g/cm^3)(100 cm/m)(980.7 $\times$ 10^{-7}N/g-force)(100^{2}cm^2/m^2) $= 883 \times 10^3$ (Tρ) Pa. Tensile strength is $0.883T\rho$ MPa.

Fabrication of Porous Foams

Natural cellular materials include sponges and wood. Foams of polymers, metals, and ceramics can be made by numerous methods. Many foams are produced by gas evolvution. Inert gasses such as CO_2 and N_2 may be dissolved under high pressure and released by decreasing the pressure. Gas bubbles may also be formed by chemical decomposition or chemical reaction. Polyurethane foam is made by reacting isocyanate with water to form CO_2. Mechanical beating also will produce foams. Foamed materials such as styrofoam can be formed by bonding together spheres that have been previously foamed.

Metallic and ceramic foams can be made from polymeric precursors. Ceramic foams can be made by dipping a polymer foam into a ceramic slurry, drying it, and then sintering it at a temperature high enough to decompose the polymer. Metallic foams can be made electroless-plating on a polymer foam and subsequent heating to drive off the polymer. Carbon foams can be made by pyrolizing polymeric foams to drive off the hydrogen.

Morphology of Foams

There are two types of foams: closed-cell foams and open-cell (or *reticulated*) foams. Open-cell foams, in which air or other fluids are free to circulate, are used for filters and as skeletons. They may be made by collapsing the walls of closed-cell foams. Closed-cell foams are much stiffer and stronger than open-cell foams because compression is partially resisted by increased air pressure inside the cells. Figure 19.2

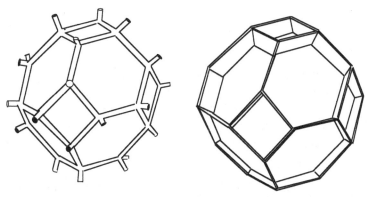

Figure 19.2. Open- and closed-cell foams modeled by tetrakaidecahedra. From W. F. Hosford, *Materials Science: An Intermediate Text*, Cambridge, 2006.

shows that the geometry of open- and closed-cell foams can modeled by Kelvin tetrakaidecahedra.

Mechanical Properties of Foams

The elastic stiffness of a foam depends on its relative density. In general the dependence of relative stiffness, E^*/E, on relative density, where E^* is the elastic modulus of the structure and E is the modulus of the solid material is of the form

$$E^*/E = (\rho^*/\rho)^n. \tag{19.1}$$

Here the densities of the structure and the material are ρ^* and ρ, respectively. For closed cells n is about 1, and for open cells n is about 2 as shown in Figure 19.3. Deformation under compression of open-cell foams is primarily by ligament bending but compression of closed-cell wall foams involves gas compression and wall stretching in addition to wall bending.

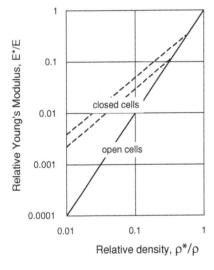

Figure 19.3. Change of elastic modulus with relative density. Data from L. J. Gibson and M. F. Ashby, *Cellular Solids: Structure and Properties*, 2nd ed., Cambridge University Press, 1999. From W. F. Hosford, *Materials Science: An Intermediate Text*, Cambridge, 2006.

Example Problem 19–2:

The deflection, δ, of a beam supported at each end and loaded with a force, F, in the center is $\delta = FL^3/(4Ebh^3)$, where E is Young's modulus, L is the span, b is the width, and h is the thickness. Consider two beams of equal L and b, one of a solid material and the other of a foam with a relative density of 0.1. Calculate the ratio of the weights of the two beams if thickness, h, is adjusted so they have the same deflection under the same force.

Solution: With the same values of F, L, δ, and b, $E_f h_f^3 = E_s h_s^3$, where the subscripts f and s refer to the foamed beam and solid beam. Solve for $h_f/\underline{h}_s = (E_s/E_f)^{1/3}$.

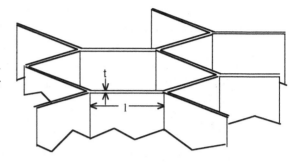

Figure 19.4. Hexagonal honeycomb structure. From W. F. Hosford, *Materials Science: An Intermediate Text*, Cambridge, 2006.

From Figure 19.3, for a relative density of 0.1, $E_f/E_s = 0.01$. Substituting $h_f/\underline{h}_s = 100^{1/3} = 4.64$. With a relative density of 0.1, the ratio of weights is $0.1(4.64) = 0.464$.

Honeycombs

Honeycombs can be made by folding and gluing thin sheets, by expanding glued sheets, by casting, and by extruding. Figure 19.4 illustrates a honeycomb structure. Panels with a high bending stiffness-to-weight ratio are often made with a honeycomb structure sandwiched between two sheets or plates. While the honeycomb doesn't directly contribute much to the stiffness, it separates the outer sheets so they have a maximum bending resistance.

Applications of Porous Materials

There are many applications of porous materials. Their capability to fill space with a minimum weight leads to their use in packaging. Life jackets and rafts use porous materials because of their low density. Examples of their use as insulators range from styrofoam cups to heat shields for spacecraft. Aluminum baseball bats are filled with foam to dampen vibrations. The low elastic moduli and high elastic strains of polyurethane foam (foam rubber) are useful in cushions and mattresses. Filters are made from porous materials.

Stiff lightweight structures such as aircraft wings are made from sandwiches of continuous sheets filled with foams or honeycombs. Open porous structures can form frameworks for infiltration by other materials, leading to the application of biocompatible implants. Open-pore structures are used as supports for catalysts.

Notes of Interest

Michael Ashby wrote, "When modern man builds large load-bearing structures, he uses dense solids; steel, concrete, glass. When nature does the same, she generally uses cellular materials; wood, bone, coral. There must be good reasons for it."

Aluminum foams can be produced by adding fine particles or oxide fibers to increase its viscosity and then injecting nitrogen or argon gas to form bubbles. An alternative is to add particles of a solid that decomposes. For example, TiH_2 added to molten aluminum produces hydrogen bubbles.

In 2006, Dr. Bryce Tappan developed metal nanofoams by igniting metal bis(tetrazolato)amine complexes. Densities as low as 11 mg/cm^3 and surface areas as high as 258 m^2/g have been achieved.

Problems

1. Calculate ρ^*/ρ for a honeycomb structure with $l/t = 10$.
2. Calculate the ratio of the weight of a solid bar to that of a foam of equal stiffness in tension if $\rho^*/\rho = 0.1$.
   a. Assume closed cells.
   b. Assume open cells.
3. In view of your answer to Problem 2, explain why beams made of foams have greater stiffness in bending than solid beams of the same weight.
4. Calculate the tensile strength in MPa of Kevlar fibers having a tenacity of 22 g force/denier.

20 Electrical Properties

Conductivity

Electrical resistivity probably varies more than any other property from one material to the next. The resistivity of diamond is about 10^{18} ohm·m and that of silver 10^{-8} ohm·m. Figure 20.1 shows this range. Conductivity, σ, is the reciprocal of resistivity, ρ.

The conductivity of a material σ, is the product of the number of charge carriers per volume, n, the charge on the carriers, zq, and their mobility, μ:

$$\sigma = nzq\mu. \qquad 20.1$$

Mobility is the ratio of drift velocity to the field and has units of $(m/s)/(v/m)$ so the units of conductivity, $ohm^{-1}·m^{-1}$, can be thought of as $(carriers/m^2)$ $(coulombs/carrier)(m^3/vs) = a/(v·m) = 1(ohm·m)^{-1}$.

If there is more than one charge carrier, the conductivity is the sum of the contributions of each carrier:

$$\sigma = \Sigma n_i zq_i \mu_i. \qquad 20.2$$

The effects of temperature, impurities, and structure on the conductivity of various materials can be understood in terms of how they influence mobility.

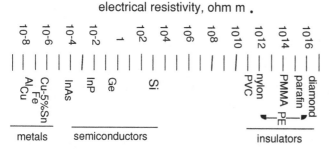

Figure 20.1. The very large range of electrical conductivity in materials.

189

Metallic Conduction

Metals are excellent conductors as shown in Table 20.1. Valence electrons are free to move throughout a metal and form an electron gas. They are in constant motion and their movement is random if there is no applied field. However, an electric field accelerates the electrons in the direction of the field, causing a net drift velocity. Any irregularity in the periodicity of the lattice will cause diffraction. The drift velocity depends on the field and the mean free path between interactions with lattice irregularities. Therefore, the conductivity and mobility are determined by the mean free path. Figure 20.2 illustrates this concept.

Table 20.1. *Electrical conductivities at 20°C*

Metal	Conductivity $(n\Omega m)^{-1}$
Aluminum	37.7
Copper	59.77
Silver	68.0
Nickel	14.6
Iron	10.5
Magnesium	22.47
Zinc	16.9
70/30 brass	16.7
Cupronickel (20% Ni)	3.9
Stainless steel (18% Cr,8% Ni	13.9

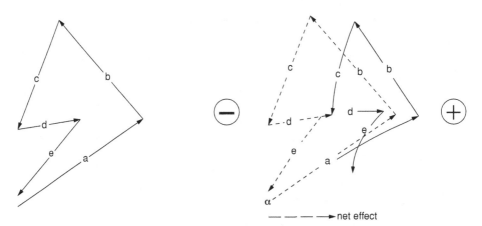

Figure 20.2. Imposition of an electrical field on the random motion of electrons creates a drift velocity. Random motion (left) and the influence of field (right).

Increased temperature makes the lattice less regular and therefore decreases the mean free path. Figure 20.3 shows that the resistivity, ρ ($= 1/\sigma$), varies linearly with temperature, except at very low temperatures.

The resistivity at a given temperature, ρ_T, can be expressed as

$$\rho_T = \rho_{273}[1 + y_T(T - 273)], \qquad\qquad 20.3$$

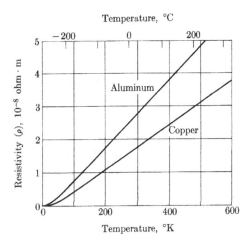

Figure 20.3. The electrical resistivities of copper and aluminum vary linearly with temperature above about 100 K. From L. H. Van Vlack, *Elements of Materials Science*, 6th ed., Addison-Wesley, 1989.

where ρ_{273} is the resistivity at 273 K and y_T is the *temperature resistivity coefficient*. For most pure metals the value of y_T is about 0.005/°C. This suggests that the mean free path and the conductivity are lowered by a factor of 1/2 between 0 and 200°C:

$$\rho_T = \rho_{273}[1 + y_T(T - 273)]. \qquad 20.4$$

Elements in solid solution also reduce the conductivity by reducing the mean free path. The increase of resistivity, $\Delta\rho_s$, caused by solutes can be expressed as

$$\Delta\rho_s = y_s X(1 - X), \qquad 20.5$$

where y_s is the *solution resistivity coefficient* and X is the mole fraction solute. In dilute solutions this simplifies to $\Delta\rho_s = y_s X$ because $1 - X$ is nearly unity. Figure 20.4 shows the effects of various solutes on the resistivity of copper.

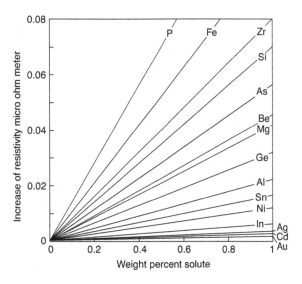

Figure 20.4. Effects of various solutes on the resistivity of copper. From W. F. Hosford, *Physical Metallurgy*, CRC, 2005.

Cold working raises the resistivity by creating vacancies and dislocations. Vacancies are more important, and the effect of each vacant lattice site is the same as that of a solute atom. Therefore, the change of resistivity caused by cold work, $\Delta\rho_{cw}$, is proportional to the number of vacancies.

The effects of temperature, solutes, and cold work are additive:

$$\rho = \rho_T + \Delta\rho_s + \Delta\rho_{cw}. \qquad\qquad 20.6$$

This is illustrated in Figure 20.5.

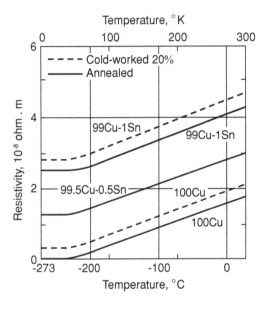

Figure 20.5. Effects of temperature, solutes, and cold work on the resistivity of copper. From C. A. Wert and R. M. Thompson, *Physics of Solids*, McGraw-Hill, 1970.

Example Problem 20–1:

Calculate the resistivity of copper containing 0.1% P at 250 °C.

Solution: Estimating $\Delta\rho_x = 0.015 \times 10^{-6}$ ohm·m from Figure 20.4 and $\rho_{Cu} = 1.5 \times 10^{-8}$ ohm·m at 250 °C from Figure 19.3, $\rho_{250} = 1.65 \times 10^{-8}$ ohm·m.

Ionic Conduction

Current is carried by movement of ions in solids under the influence of an electric field, just as in liquid solutions. The difference is that the rate of movement of ions in solids is much slower than in liquids. The charge on each ion is the product of the valence and the charge of an electron. The mobility is

$$\mu = \frac{zqD}{kT}, \qquad\qquad 20.7$$

where D is the diffusivity, k is Boltzmann's constant, and T is the absolute temperature. Because D increases exponentially with temperature, $D = D_0 \exp(-Q/RT)$, the ionic conductivity, σ_i, increases with temperature:

$$\sigma_i = \frac{nz^2q^2D_0}{kT} \exp(-Q/RT).$$ 20.8

The effect of impurities in solid solution is to increase the number of carriers so the conductivity can be expressed as the sum of the intrinsic conductivity of the pure material, σ_{in}, and the extrinsic conductivity caused by the impurity, σ_{ex},

$$\sigma = \sigma_{in} + \sigma_{ex}.$$ 20.9

Increases of both temperature and impurities content increase conduction in ionic crystals.

Energy Bands

The difference among metals, semiconductors, and insulators can be understood in terms of the energy levels of the outer electrons. Electrons in isolated atoms can exist only in discrete energy levels, with no more than two electrons occupying any level. However, as a large number of atoms are brought together to form a crystal, the electric fields of the various atoms interact causing permissible energy levels to split into bands. The number of electrons in each energy band can be no more than twice the number of atoms. Conductivity depends on whether the permissible energy levels overlap, are only partially filled, are completely filled, or are separated by an energy gap between the bands. Figure 20.6 illustrates these possibilities schematically. With metals either the valence band overlaps the conduction band or it is only partially filled so there is a very large number of carriers. In semiconductors the valence band

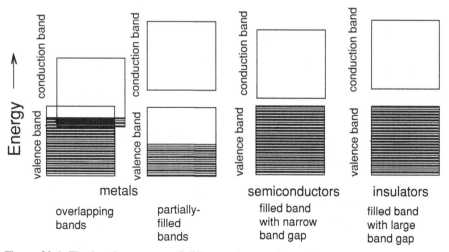

Figure 20.6. The band structures of different classes of materials.

is filled, but there is only a small energy gap between the valence and conduction bands. With insulators the valence band is filled, and there is a large energy gap between the valence and conduction bands.

Intrinsic Semiconduction

The energy gaps of several semiconductors are listed in Table 20.2, so with thermal agitation some small number of electrons, n, will have energies high enough to move into the conduction band as sketched in Figure 20.7. For a pure (*intrinsic*) semiconductor the number, n, depends on temperature:

$$n = A\exp(-E_g/9.102kT). \hspace{3cm} 20.10$$

Table 20.2. *Energy gaps and mobilities of intrinsic semiconductors*

Semiconductor	Energy gap (eV)	Mobility (m²/V·s) electrons	holes
Silicon	1.1	0.19	0.425
Germanium	0.7	0.36	0.23
GaAs	0.7	0.60	0.08
InAs	0.36	2.26	0.26
InP	1.3	0.47	0.015

Figure 20.8 illustrates this on an atomic level. For every electron promoted to the conduction band, a hole is generated in the valence band. The electrons in the

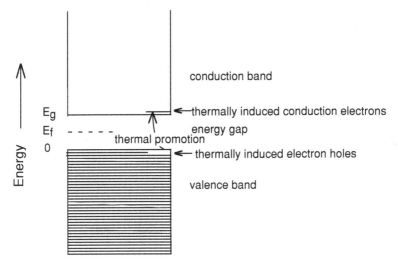

Figure 20.7. Band structure of an intrinsic semiconductor. A few electrons are thermally promoted from the valence band to the conduction band.

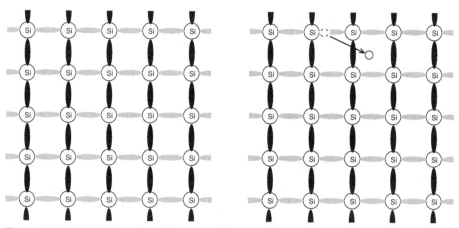

Figure 20.8. Each silicon atom shares two electrons with each of four near neighbors (left). If an electron is removed from bonding, a conduction electron and an electron hole are created.

conduction band are carriers of negative charge and the holes in the valence band act as carriers of positive charge. With equal numbers, $n_n = n_p$; so

$$n = n_n + n_p, \qquad\qquad 20.11$$

where n_n and n_p are the numbers of conduction electrons (negative carriers) and holes (positive carriers). Both electrons and electron holes contribute to the conductivity. In general,

$$\sigma = n_n q \mu_n + n_p q \mu_p, \qquad\qquad 20.12$$

where μ_n and μ_p are the mobilities of electrons and holes. For a pure (intrinsic) semiconductor, $n_n = n_p = n/2$ so

$$\sigma = (n/2)q\mu_n + (n/2)q\mu_p = (n/2)q(\mu_n + \mu_p). \qquad\qquad 20.13$$

Because n is exponentially related to temperature,

$$\sigma = \sigma_0 \exp(-E_g/2kT). \qquad\qquad 20.14$$

Figure 20.9 shows that the conductivity of germanium increases exponentially with temperature.

Example Problem 20–2:

The conductivity of silicon at room temperature is $4 \times 10^{-4} (\text{ohm·m})^{-1}$. Calculate the number of carriers, $n = p$.

Solution: Since $\sigma = nq(\mu_e + \mu_h)$, $n = \sigma/[q(\mu_e + \mu_h)] = 4 \times 10^{-4}/[(1.6 \times 10^{-19}) (0.19 + 0.425)] = 1.33 \times 10^{16}/\text{m}^3$.

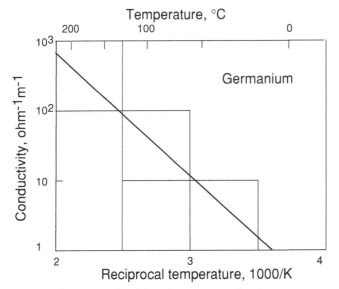

Figure 20.9. The conductivity of pure germanium increases exponentially with temperature.

Example Problem 20–3:

Knowing that the conductivity of pure silicon is $4 \times 10^{-4}(\text{ohm·m})^{-1}$ at 20 °C, calculate the conductivity at 200 °C.

Solution: From equation 20.14, $\sigma_2/\sigma_1 = \exp[(-1.1 \text{ eV}/2 \times 86.1 \times 10^{-6})(1/473 - 1/293)] = 4.01 \times 10^3$, so $\sigma_2 = (4.01 \times 10^3)(4 \times 10^{-4}) = 1.6 \,(\text{ohm·m})^{-1}$.

The energy gap is so large in insulators that almost no electrons are promoted to the conduction band.

Extrinsic Semiconduction

Addition of even very small amounts of impurities greatly alters the conductivity. For example, impurities of group V elements (N, P, As, Sb) add an extra electron, which can act as a conductor. The extra electron is in a donor state with an energy level only slightly below that of the conduction band as illustrated in Figure 20.10. Very little thermal energy $(E_g - E_d)$ is required to promote an electron from the donor state to the conduction band. At room temperature almost all of the donor electrons are promoted. In this case the conductivity is primarily by donor electrons and n_n is approximately the number of group V atoms per volume and the semiconductor is *n type* (negative carriers). Figure 20.11 shows the temperature dependence in this case.

The effect of group III (Al, Ga, In) impurities is similar. It takes very little thermal energy to promote an electron from the valence band to the acceptor state.

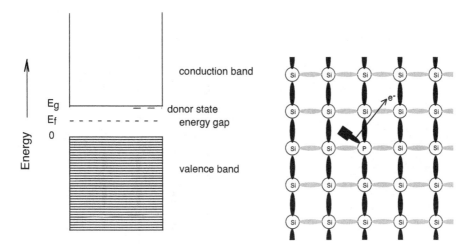

Figure 20.10. Group V impurities create donor. The energy level of donors is close to the conduction band (left). The extra electron can easily be promoted to conduction (right).

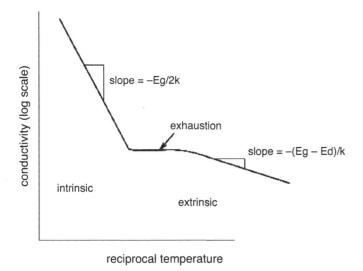

reciprocal temperature

Figure 20.11. The temperature dependence of conductivity in an extrinsic semiconductor. At very low temperatures, conductivity depends on thermal excitation of donor or acceptor levels. Near room temperature there is saturation, and at high temperatures intrinsic conduction prevails.

This creates a hole in the valence band (Figure 20.12). At room temperature the acceptor state is almost filled so the number of electron holes is approximately the number of group III atoms per volume and the semiconductor is *p type* (positive carriers).

In the exhaustion range, the conductivity is approximately proportional to the impurity concentration.

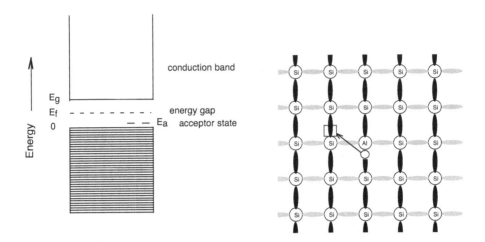

Figure 20.12. Group III impurities create electron holes. The acceptor energy level is slightly above the valence band (left). The missing electron is easily promoted to an electron hole.

Example Problem 20–4:

Calculate the number of carriers/m^3 if 1 part per million (by weight) phosphorus were added to silicon.

Solution: $(10^{-9} \text{ gP/gSi})(6.023 \times 1023 \text{ atoms/mole})/(30.97 \text{ g/mole}) = 1.945 \times 10^{14}$ atoms P per gram Si. $(1.945 \times 10^{14} \text{ atoms P per gram Si})(2.33 \text{ MgSi/m}^3) = 4.53 \times 10^{20}$ atoms P/m$^3 = n_e (n_h \approx 0)$.

III-V Semiconductors

Compounds of group III and group V elements are semiconductors. Examples include AlSb, GaP, GaAs, InP, and InSb. Zinc sulfide (groups II-VI) is also a semiconductor. Either nonstoichiometric compositions or foreign impurities create n- or p-type semiconductors. Figure 20.13 illustrates schematically the structure of these compounds.

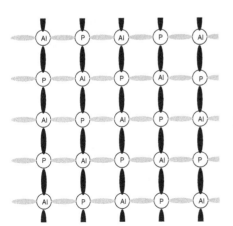

Figure 20.13. Structure of a group III-V semiconductor.

Table 20.3 lists some of the properties of semiconductors.

Table 20.3. *Properties of some semiconductors*

Material	Energy gap, E_g		Mobility, m²/vs		Intrinsic conductivity	Lattice parameter
	10^{-18} J	eV	electrons	holes		
			μ_n	μ_n	(ohm·m)$^{-1}$	nm
Silicon	0.176	1.1	0.19	0.0425	5×10^{-16}	0.543
Germanium	0.112	0.7	0.36	0.23	2	0.566
AlSb	0.26	1.6	0.02			0.613
GaP	0.37	2.3	0.019	0.012		0.545
GaAs	0.22	1.4	0.88	0.04	10^{-6}	0.565
InP	0.21	1.3	0.47	0.015	500	0.587
InAs	0.058	0.36	2.26	0.026	10^{-4}	0.604
InSb	0.029	0.18	8.2	0.17		0.648
ZnS	0.59	3.7	0.014	0.0005		

n-p Rectifiers

A *rectifier* or *diode* allows current to pass in only one direction. A p-n junction can act as a rectifier. Figure 20.14 shows that if a forward bias is applied to the junction, holes in the p-type material and electrons in the n-type material will flow to the junction where they will combine, allowing current to flow. More electrons and more holes are created where external leads are connected. With a reverse bias, electrons and holes move away from the junction, creating a zone depleted of carriers.

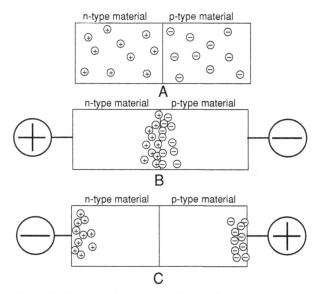

Figure 20.14. A p-n junction rectifier has both p- and n-type regions (A). With forward bias (B) electrons and holes both move toward the junction, allowing current to pass. With a negative bias (C) the region near the junction becomes depleted in carriers so no current can flow.

Transistors

There are two types of transistors. Junction transistors are either of the n-p-n type or the p-n-p type. In a p-n-p transistor (Figure 20.15), the n-type material forms a thin base. Because the base is so thin, holes in the emitter can cross into the collector. A small input voltage signal controls the number of such holes and therefore the collector current. The voltage drop across the large resistor in the collector circuit amplifies the smaller input voltage signal.

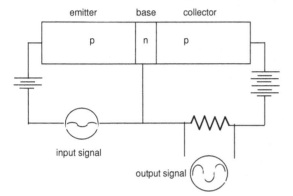

Figure 20.15. A p-n-p transistor. A small input signal imposed on the emitter causes holes to penetrate the base and allows current in the collector. Because the voltage across the collector is high, the output signal is at high voltage.

The other type of transistor is the metal oxide semiconductor field effect transistor (MOSFET) sketched in Figure 20.16. The gate, made from polycrystalline silicon, is separated from the channel by layer of SiO_2. A small signal voltage between the source and gate creates an electric field that penetrates the channel and controls its conductivity. Modulating the field between the source and gate will modulate the conduction between the source and drain.

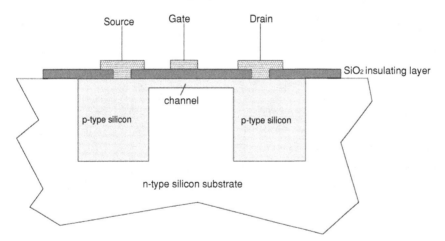

Figure 20.16. A MOSFET.

Dielectric Properties

Imposition of an electric field, E, between two parallel plates develops a charge density, D, on them

$$D = \varepsilon E, \qquad\qquad 20.15$$

where ε is the electrical permittivity of the material between the plates. The units of charge density, D, are coul/m^2 and the units of electric field, E, are volts/m, so the units of permittivity are (coul/m^2)/(volts/m) = farads/m. The dielectric constant, κ, is given by

$$\kappa = \varepsilon/\varepsilon_o, \qquad\qquad 20.16$$

where ε_o = 8.854 farads/m is the permittivity of vacuum.

An insulator in an electric field becomes polarized. It develops an interior field in response to the imposed field. There are several ways this may happen. One is electronic polarization of the electron fields around an atom. As suggested in Figure 20.17, electrons may be attracted to the positive side of the field.

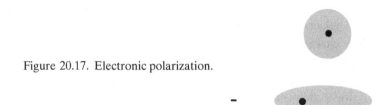

Figure 20.17. Electronic polarization.

Ions may be displaced from their normal lattice positions by a field as illustrated in Figure 20.18.

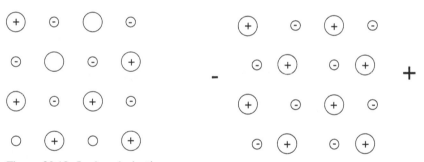

Figure 20.18. Ionic polarization.

Molecules with internal fields may become oriented in the presence of a field as shown in Figure 20.19.

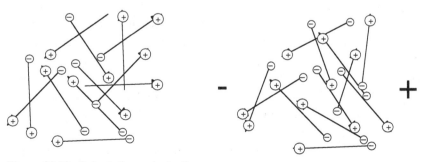

Figure 20.19. Orientation polarization.

Finally ions may migrate to new positions under the influence of a field as illustrated in Figure 20.20.

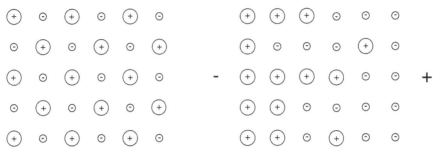

Figure 20.20. Space charge polarization.

Dielectric constants are often frequency dependent because orientation and space charge polarization require time for molecules and ions to move. Figure 20.21 shows the frequency dependence of polyvinylchloride (PVC) and polytetrafluroethylene (PTFE). Because PTFE is symmetric, there is no orientation polarization. In contrast, the position of a chlorine atom in PVC leads to orientation polarization, which requires time to adjust to a field.

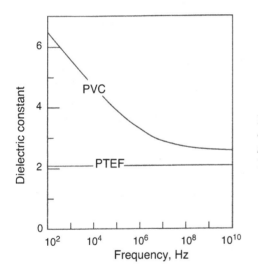

Figure 20.21. Frequency dependence of polarization of PVC and PTFE. Data from L. H. Van Vlack, *Elements of Materials Science*, 6th ed., Addison-Wesley, 1989.

Some polymer molecules are polar. Their electrons are not shared equally between bonding atoms. Atoms like fluorine, oxygen, and nitrogen exert a greater force on the electron cloud than other atoms, with the result that the electron cloud is off center.

The same is true of some crystals. One example is quartz. Another is barium titanate. Above $120\,°C$, $BaTiO_3$ is cubic with the Ti^{4+} ion in the center of a cubic unit cell, O^{2-} ions in face-centering positions, and Ba^{2+} ions at the corners. However, at lower temperatures the cell becomes tetragonal with the crystal structure shown in Figure 20.22. The Ti^{4+} ion shifts in one direction and the O^{2-} ions in the opposite direction relative to the Ba^{2+} ions at the corners. These displacements of the centers of positive and negative charge set up a permanent dipole. The polarization is given by

$$P = \Sigma[Qd/V], \qquad\qquad 20.17$$

where Q is the charge on an ion, d is its displacement from the center of charge, and V is the volume.

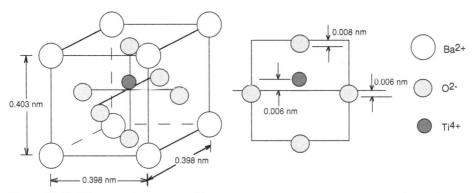

Figure 20.22. The crystal structure of barium titanate. Note that the centers of negative and positive charges are separated, creating a permanent dipole.

Electrode and the negative ions are pulled toward the positive electrode. There is no voltage in the absence of an external field because the accumulated charge just balances the dipole.

Piezoelectric Behavior

Piezoelectricity is the capability of some crystals to generate a voltage in response to applied stress and conversely to undergo a reversible strain in response to a voltage. The word piezoelectricity comes from the Greek *piezein*, which means to squeeze or press. The effect finds useful applications such as the production and detection of sound, generation of high voltages, electronic frequency generation, sonar, and ultra-fine focusing of optical assemblies.

An applied external field distorts the material as indicated in Figure 20.23 as the internal negative charge is attracted toward the additional positive voltage. Conversely an external force, which strains the crystal, produces a voltage.

Example Problem 20–5:

Using equation 20.16, calculate the polarization of $BaTiO_3$.

Solution:

$$P = [4(0.006) + 4(-2)(-0.006) + 2(-2)(-.008)]$$
$$(0.1602 \times 10^{-18})/[(0.403 \times 10^{-9})(0.398 \times 10^{-9})^2] = 0.16 \, coul/m^3.$$

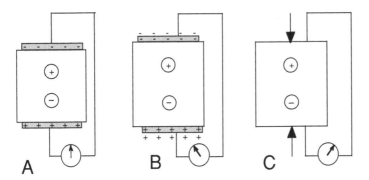

Figure 20.23. A piezoelectric crystal (A) is distorted by an electric field (B) and an external force produces a voltage (C).

The polarization changes as a crystal is strained elastically. Since neither Q nor V change appreciably with strain, the change of polarization is $dP/P = dd/d = d\varepsilon$.

Example Problem 20-6:

Knowing that Young's modulus for $BaTiO_3$ is 72 GPa, calculate the stress necessary to change the polarization by 0.5%.

Solution: $dP/P = dd/d = d\varepsilon = 0.005 = \sigma/E, \sigma = E(dP/P) = 72\,\text{GPa}(0.005) = 360\,\text{MPa}$.

Piezoelectric materials are useful in applications where mechanical vibrations need to be developed by vibrating fields, such as for sending units of sonar, in audio speaker systems, in analog watches, and for high-frequency sound generators. They are also used where mechanical vibrations are used to produce electrical signals as in the receiving units of sonar, microphones, and frequency standards. Above 120 °C (its Curie temperature), barium titanate has a cubic crystal structure that is not piezoelectric. PZT [Pb(Zr,Ti)O₃] is piezoelectric up to 480 °C. Lead meta-niobdate has a Curie temperature of 570 °C. PZT is the most widely used piezoelectric material.

Vibrating quartz crystals are often used for frequency standards. Their natural vibrational frequency depends on the crystal dimensions. Once a crystal is cut to its final dimensions, its frequency is constant to one part in 10^8. Piezoelectric crystals are used as pressure sensors.

Notes of Interest

Although three patents for the field effect transistor were issued in 1928 to J. E. Lilienfeld in Germany, they were ignored by industry. In 1934, Oskar Heil patented another field effect transistor. In 1947, W. Shockley, J. Bardeen, and W. H. Brattain made the first practical transistor at Bell Labs. This work followed their wartime efforts to produce extremely pure germanium for diodes used in radar. Bell Labs didn't publicly announce the transistor until 1948. A French development of the transistron is considered to be independent. Mataré had first observed transconductance effects during the manufacture of germanium duodiodes for German radar equipment during World War II.

The piezoelectric effect was discovered by Pierre and Jacques Curie in 1880. However, it was of no practical use until the 1960s because of the weak signals. In 1950, the charge amplifier principle was patented by W. P. Kistler, which allowed practical use of this effect.

Problems

1. Predict the conductivity of an alloy of copper containing 2% nickel at 250 °C.
2. Calculate Y_T for aluminum from Figure 20.3.
3. Germanium is doped with 1 part per million (by weight) aluminum.
   a. Is this an intrinsic, n-type, or p-type semiconductor?
   b. Calculate the number of carriers per volume, n.
   c. Predict the conductivity at 50 °C.
4. Which would have a higher dielectric constant:
   a. polyethylene or polyvinyl acetate?
   b. PVC at 25 °C or PVC at 100 °C?
   c. PVC at 10 Hz or PVC at 10^6 Hz?
5. A compound AB has the structure of NaCl with four A^{2+} ions and four B^{2-} ions per unit cell. The dimensions of the unit cell are 0.50 nm. The mobility of the A^{2+} ions is 2.0×10^{-15} m^2/V. Calculate the electrical resistivity.
6. Does adding 1% silver to copper increase its conductivity? If so, why? If not, why not?
7. If there were more As vacancies than Ga vacancies in GaAs, which type of semiconductor would result?
   a. p-type
   b. n-type
   c. intrinsic semiconductor
8. Determine $\Delta \rho_s$ for 20 weight % nickel in copper.

21 Optical and Thermal Properties

Spectrum of Electromagnetic Radiation

There is a very wide spectrum of electromagnetic radiation as shown in Figure 21.1. The visible portion of the spectrum is enlarged in Figure 21.2.

The wavelength and frequency are related:

$$\lambda = h/\nu. \tag{21.1}$$

The energy of light is given by

$$E = hc/\lambda, \tag{21.2}$$

where λ is the wavelength, c the velocity of light (3×10^8 m/s), and h is Planck's constant (6.62×10^{-27} erg·s). With λ expressed in μm, the energy in eV is $E = (1.24 \times 10^3/\lambda)$ eV.

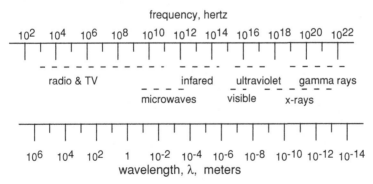

Figure 21.1. The electromagnetic spectrum.

Example Problem 21–1:

Calculate the energy of the photons of infrared light ($\lambda = 1200$ nm), yellowish green light ($\lambda = 550$ nm), and ultraviolet light ($\lambda = 200$ nm).

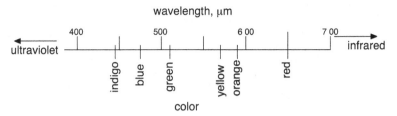

Figure 21.2. The visible spectrum.

Solution: For the infrared, $E = 1.24 \times 10^3/1200 = 1.03\,\text{eV}$; for the yellowish green, $E = 1.24 \times 10^3/550 = 2.25\,\text{cV}$; and for the ultraviolet, $E = 1.24 \times 10^3/200 = 6.2\,\text{eV}$.

Index of Refraction

The index of refraction, n, of a material is the ratio of the velocity of light in a vacuum, v_o, to that in the material: v_m:

$$n = v_\text{o}/v_\text{m}. \qquad 21.3$$

When light passes from one material into another, it is refracted. Snell's law states that the angles to the surface normal, ϕ, are given by

$$\sin\phi_2/\sin\phi_1 = v_2/v_1 = n_1/n_2, \qquad 21.4$$

as illustrated in Figure 21.3.

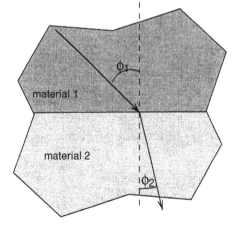

Figure 21.3. Refraction of light: $n_1 < n_2$ so $\phi_2 < \phi_1$.

The index of refraction is anisotropic in non-cubic crystals. This is the cause of birefringence or double refraction.

Table 21.1 gives the indices of refraction for several materials. Two indices are listed for quartz and calcite because they are birefringent.

Table 21.1. *Indices of refraction of several materials*

Material	Index of Refraction, n (20°C. $\lambda = 590$ nm)
Vacuum	1.0 (by definition)
Air	1.0003
Water	1.333
Optical glass (heavy flint)	1.650
Optical glass (crown)	1.517
Quartz	1.544–1.553
Calcite	1.658–1.486
Diamond	2.438
PTFE	1.4
PP	1.47
PMMA	1.49
Polystyrene	1.60

Example Problem 21–2:

If one perceives that an object in the water lies at an angle of 45°, what is the true angle θ in Figure 21.4?

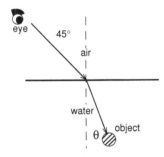

Figure 21.4. Appearance of an object in the water.

Solution: From equation 21.4, $\sin \theta = \sin 45 (n_{air}/n_{water}) = 0.707(1.0003/1.333) = 0.551; \theta = 32°$.

The index of refraction increases with the electronic polarization and therefore is frequency dependent. This is why a prism separates light according to wavelength. This is illustrated schematically in Figure 21.5.

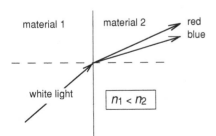

Figure 21.5. Different diffraction of different wavelengths.

Reflectivity

The reflectivity, R, or fraction of the light that is reflected is directly related to n:

$$R = [(n - 1)/(n + 1)]^2. \qquad 21.5$$

For example, diamond has a much higher reflectivity than quartz.

Absorption

As light passes through a material, a fraction of it is absorbed. The intensity, I, decreases with distance, x:

$$dI = -\alpha \, dx, \qquad 21.6$$

where α is the absorption coefficient. After integrating,

$$I = I_o \exp(-\alpha x). \qquad 21.7$$

Example Problem 21–3:

If 80% of the light is transmitted through a plate 2 cm thick, what fraction of the light will be transmitted through a 10 cm plate of the same material?

Solution: Rearranging equation 21.7 and solving for α, $\alpha = -(1/x)\ln(I/I_o) = (1/2)\ln(0.8) = 0.112/\text{cm}$. For 10 cm, $I/I_o = \exp[-(10)(0.002)] = 32.1\%$.

The absorption mechanisms correspond to the polarization mechanisms. Figure 21.6 illustrates schematically the frequency ranges of each mechanism. For visible light, electronic polarization is most important.

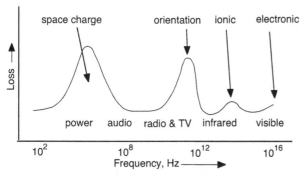

Figure 21.6. Absorption mechanisms.

Absorption depends on the presence of impurities or second phases. In insulators, impurity absorption occurs if the incident radiation has an energy level that can excite electrons from the valence band into the acceptor level or from the donor level into the conduction band. Different impurities lead to different absorption spectra and hence different colors in transparent materials. Figure 21.7 shows the transmitted spectrum of green glass.

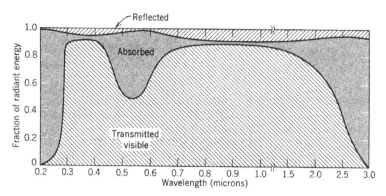

Figure 21.7. Spectrum of light incident on green glass.

The fraction of light transmitted through a material, T, plus the fraction absorbed, A, and the fraction reflected, R, equals unity:

$$T + A + R = 1. \qquad\qquad 21.8$$

Opacity

There are two reasons why a material may be opaque. One is that photons may be absorbed by raising the energy level of electrons. This is why metals are opaque. As soon as the electron is promoted to a higher energy level, it drops back down to a lower level, emitting a photon, thus reflecting light. A second reason is that there may be internal surfaces that reflect light. Although a single crystal of salt is transparent, a pile of table salt is opaque because light is refracted at air-crystal interfaces every time it enters or leaves a crystal. Most ceramics contain pores, and are therefore opaque. Even if there is no porosity, polycrystals can be opaque if the velocity of light varies with crystallographic direction.

There are no internal interfaces in polymers that are completely amorphous (e.g., PMMA). In polyethylene there are both amorphous and crystalline regions, so polyethylene is opaque. The crystallinity is increased by stretching so it becomes more transparent.

Luminescence

When the energy of an electron is excited to a higher energy level and then falls to a lower one, electromagnetic radiation is emitted. The radiation is in the visible range if the difference in energy states is between 1.8 and 3.1 eV. The energy source to excite the electron to the higher energy level may be higher energy electromagnetic radiation such as ultraviolet light, electron bombardment, or chemical (such as the light of fireflies). If re-emission of the light occurs in less than one second after stimulation, the phenomenon is called *fluorescence*. If the time delay is longer, it is called *phosphorescence*.

Light-emitting Diodes (LEDs)

When current is passed through a p-n junction diode, electrons drop into holes. The decrease of the energy associated with the electronic states is converted to electromagnetic radiation. This phenomenon is called *electroluminescence*. The color of the light emitted from *light-emitting diodes* (LEDs) depends on the drop energy difference of electrons in the p and n parts of the junction. The first LEDs were of gallium arsenide and emitted red light. Some of the other materials used for LEDs are aluminum gallium arsenide (red and infrared), aluminum gallium phosphide (green), gallium phosphide (red, yellow, and green), and many others.

There are also organic LEDs made of small molecules or polymers. If polymeric, they have the advantage of being flexible and have potential use in cloth and wall decorations.

Photovoltaic Cells

Photoelectric cells, including solar cells, convert the energy of light into electrical energy. When light of sufficient energy strikes a semiconductor, valence electrons are promoted to the conduction band in n-type material and holes in the valence band are generated in p-type material. A voltage difference is set up at a p-n junction. This resulting electric current can be stored in a battery. Single crystals of silicon are most widely used but cells made from polycrystalline silicon are cheaper.

Lasers

The word laser is an acronym for light amplification by stimulated emission of radiation. For example, a ruby laser is made from a crystal of aluminum oxide containing a small amount of chromium oxide. The crystal ends are ground flat and one is heavily silvered so it is completely reflective. The other end is lightly silvered so as to be partially reflective. If the crystal is subjected to radiation from xenon gas ($\lambda = 560$ nm), the electrons in most of the chromium ions are excited to a higher metastable energy state. After a few milliseconds, they drop to their ground state, emitting a beam of coherent red light with a wavelength of 694.3 nm.

YAG (yttrium-aluminum-garnet) lasers contain Nd^{3+} ions and emit light of 1065 nm wavelength. They emit a continuous light beam of more than 1 kW power or can be pulsed with pulses as short as 10^{-12} s with a power as great as 10^9 kW.

Fiber Optics

Very-high-frequency signals can be transmitted by light carried in glass fibers. The losses in optical fibers are minimized with fibers in which the index of refraction continuously decreases between the axis and the cladding so the light rays bend gradually as they approach the cladding, rather than reflecting abruptly from the core-cladding boundary. The resulting curved paths reduce multipath dispersion. A parabolic variation of the index of refraction with distance from the center is ideal.

Table 21.2. *Melting points, thermal expansion coefficients, and Young's moduli for several materials*

Material	Melting point (°C)	Coefficient of thermal Expansion [°C^{-1}($\times 10^{-6}$)]	Young's modulus (GPa)
Lead	327	29	14
Magnesium	649	25	45
Aluminum	660	22.5	70
Copper	1084	17	110
Iron	1538	11.7	205
Molybdenum	2620	4.9	315
Tungsten	3410	4.6	350
Polyethylene	120	180	$\approx$0.2
Nylon	266	100	2.8
MgO	2800	9	205
Al$_2$O$_3$	2050	9	350
TiC	3140	7	350

Optical fibers are made from large-diameter silica preformed rods with controlled refractive index profiles that are established by chemical vapor deposition. Typically the difference between core and cladding is less than 1%. The preformed rods are pulled to form the long, thin fibers.

Thermal Expansion

Almost all materials expand when heated. The fractional increase in dimension per degree temperature change is called the linear coefficient of thermal expansion. The coefficient of thermal expansion of a material is inversely related to its melting point. Young's modulus and melting point increase with the strength of bonds between atoms or molecules.

Table 21.2 lists the coefficient of thermal expansion along with the melting points and Young's modulus for several materials. Figure 21.8 shows the inverse correlation of thermal expansion with melting point.

The fractional volume expansion is three times the linear expansion; however, there are exceptions. An alloy of 64% Fe, 36% Ni has nearly zero coefficient of expansion.

Thermal Conductivity

In metals heat conduction is primarily by movement of free electrons. For this reason, thermal and electrical conduction are closely related. The ratio of thermal and electrical conductivities of most metals and alloys is about 7×10^{-6} W·ohm/K at room temperature if the thermal conductivity, k, is in W/m^2/(K/m) and the electrical conductivity is expressed in (ohm·m)$^{-1}$. This is known as the Wiedemann-Franz ratio. The correlation is shown in Figure 21.9.

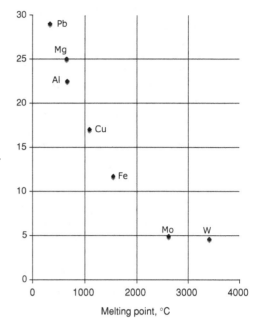

Figure 21.8. Inverse correlation of thermal expansion with melting point.

The thermal conductivities of non-metals are much lower and do not correlate with the electrical conductivities. In non-metals, thermal conduction is largely by lattice vibrations. The thermal conductivities of polymers and ceramics generally correlate with the elastic moduli. Diamond, which has the highest elastic modulus of all materials, also has the highest thermal conductivity. The thermal conductivities of most ceramics are much lower than those of metals but are one or two orders of magnitude greater than those of polymers.

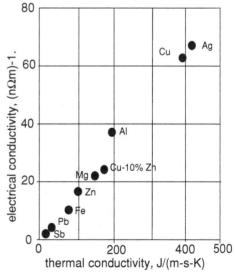

Figure 21.9. Correlation between thermal and electrical conductivities in metals.

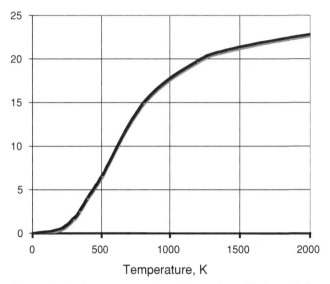

Figure 21.10. Temperature dependence of specific heat of diamond.

Specific Heats

The specific heat, C_p, is the energy required to raise the temperature of a material by one degree. The specific heat of solids at room temperature is nearly $3kT/2$ per atom, which is 24.9 J/mol·K. This is known as *Dulong and Petite's law*. Below room temperature the specific heat of solids decreases with decreasing temperature to zero at 0 K (Figure 21.10). At very low temperatures, C_p is proportional to the cube of the absolute temperature.

Latent Heats of Transformation

There is a latent heat of transformation for every phase change. For metals, the latent heat of fusion is

$$\Delta H_f \approx 10 T_m \text{ MJ}/(\text{mol} \cdot \text{K}).$$
                                                                                            21.9

Latent heats associated with transformation of crystal structure in the solid are much lower. For example, the latent heat associated with the $\alpha \rightarrow \gamma$ transformation in iron is less than a tenth that of fusion.

Thermocouples

The number of high-energy electrons in a metal increases with temperature so there is a voltage gradient across a metal that is hot on one end and cool on the other. The voltage is different for different metals. If wires of two different metals are connected and one end heated, a voltage will be generated across the cold ends. By measuring this voltage, one can determine the temperature of the hot end. Common thermocouple pairs are copper vs. constantan (60%Cu, 40% Ni), chromel (90% Ni,

10% Cr) vs. alumel (94% Ni, 2% Al, 3% Mn, 1% Si), and platinum versus platinum containing 10% rhodium.

Notes of Interest

From 1670 to 1672, Sir Isaac Newton lectured on optics. During this period he investigated the refraction of light, showing that white light could be decomposed into a spectrum of colors by a prism. He also showed that with a second prism the colored light would combine to form white light. Newton argued that light is composed of particles, but he could explain diffraction only by associating them with waves. In 1704 Newton wrote *Optics* in which he expounded his corpuscular theory of light. He considered light to be made up of extremely subtle corpuscles, and that ordinary matter was made of grosser corpuscles.

Although LED traffic lights are brighter, use less energy, and have much longer lives than the old incandescent lights, they do have one disadvantage for use in northern states. They don't emit enough heat to melt snow.

Problems

1. A material absorbs radiation with a wavelength less than 15 μm but transmits most of the radiation with greater wavelengths.
   a. Determine its energy gap.
   b. Is the material likely to be a metal, polymer, or semiconductor?
2. 1 mm of a material transmits 90% of the light incident on it. What fraction of the light would be transmitted by a thickness of 10 mm?
3. At what angle will light in water not be transmitted into air? The indices of refraction for air and water are 1.00028 and 1.333, respectively. See Figure 21.11.
4. The electrical resistivity of monel (67% Ni, 33% Cu) at room temperature is 48.2 $\mu\Omega$·cm. Estimate its thermal conductivity.
5. Estimate the heat of fusion in J/kg for aluminum.
6. The specific heat of iron at 20 K is 0.0046 cal/(g·K). Estimate the specific heat at 10 K.
7. A bimetallic strip is made by laminating strips of steel to copper. As the strip is heated, will it bend with the copper on the inside, outside, or will it remain straight?

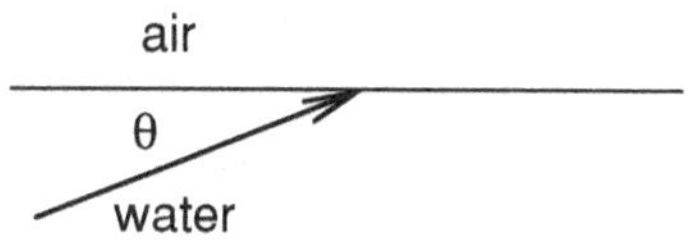

Figure 21.11. Critical angle for refraction of light at an air-water interface.

Magnetic Materials

Ferromagnetism

All materials have some interaction with magnetic fields. However, the interaction is strong only in *ferromagnetic* materials. In this chapter, the term *magnetic* will mean ferromagnetic. Iron, nickel, and cobalt are the only elements that are magnetic at room temperature, although manganese may act magnetically in some alloys. Several rare earth elements are magnetic at temperatures below room temperature and are useful in certain magnetic alloys. Magnetic materials may be broadly classified as either soft or hard. In soft magnetic materials, the direction of magnetization is easily reversed. These are used in transformers, motors, generators, solenoids, relays, speakers, and electromagnets for separating scrap. Hard magnetic materials are those in which it is difficult to change the direction of magnetization. Uses of permanent magnets include compasses, starter motors, antilock brakes, motors, microphones, speakers, disc drives, and frictionless bearings.

Domains

A magnetic material consists of magnetic domains in which the directions of unbalanced electron spins of individual atoms are aligned with each other. Magnetization is a result of these electron spins. In a material that appears to be not magnetized, the fields of the domains are arranged so that their fields cancel as illustrated in Figure 22.1. The magnetization within the domains is in a crystallographic direction of easy magnetization.

When an external magnetic field is applied, the domains in closest alignment with that field grow as illustrated in Figure 22.2. This causes the magnetization to increase as shown in Figure 22.3. At high fields the direction of magnetization rotates out of the easy crystallographic direction to be aligned with the external field.

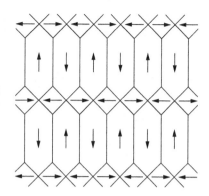

Figure 22.1. Domains in an "unmagnetized" material forming closed magnetic circuits. From W. F. Hosford, *Physical Metallurgy*, CRC, 2005.

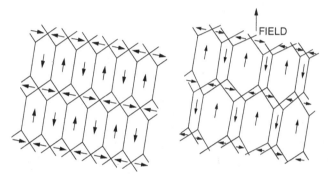

Figure 22.2. Growth of favorably oriented domains in a magnetic field. From W. F. Hosford, *ibid.*

Figure 22.3. *B-H* curve showing domain changes under increasing fields. *H* is the imposed magnetic field and *B* is the material's response to the field. From W. F. Hosford, *Physical Metallurgy*, CRC, 2006.

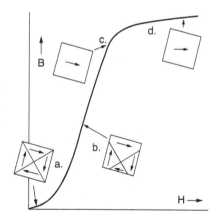

Saturation Magnetization

The magnetic moment set up by one unbalanced electron spin is called a *Bohr magneton*. One Bohr magneton equals 9.27×10^{-24} A $\cdot$ m^2. Table 22.1 lists the magnetic moments of the transition elements in the metallic (un-ionized) state.

Table 22.1. *Bohr magneton associated with several methods*

Element	2s electrons	3d electrons	Unbalanced 3d electrons (Bohr magnetons)
Sc	2	1	1
Ti	2	2	2
V	2	3	3
Cr	1	5	5
Mn	2	5	5
Fe	2	6	4
Co	2	7	3
Ni	2	8	2
Cu	1	10	0

Example Problem 22–1:

Calculate the saturation induction of iron.

Solution: The magnetic moment of an iron atom is 4 Bohr magnetons, so the magnetic moment is $(4 \text{ Bohr magneton/atom})(9.27 \times 10^{-24} \text{ A} \cdot \text{m}^2/\text{Bohr magneton}) = 37.1 \times 10^{-24} \text{ A} \cdot \text{m}^2/\text{atom}$. Iron has a density of 7.87 Mg/m^3 and the atomic weight of iron is 55.85 g/mole so the number of Fe atoms per volume is $(7.87 \times 10^6 \text{ g/m}^3)(6.02 \times 10^{23} \text{ atoms/mol})/(55.85 \text{ g/mole}) = 8.5 \times 10^{28} \text{ Fe atoms/m}^3$. At saturation, the moment is $(37.1 \times 10^{-24} \text{ A} \cdot \text{m}^2/\text{atom})(8.5 \times 10^{28} \text{ Fe atoms/m}^3) = 3.15 \times 10^6 \text{ A/m}$. $B_s = (4\pi \times 10^{-7} \text{V} \cdot \text{s/A} \cdot \text{m})(3.15 \times 10^6 \text{ A/m}) = 3.96$ tesla.

Note the experimental value is about 2.1 tesla. In the metallic state not all of the magnetic moments of atoms are aligned. In ceramics, however, the alignment is nearly perfect.

Ceramic Magnets

In magnetite (Fe_3O_4) the oxygen ions form a face-centered cubic lattice as shown in Figure 22.4. In the unit cell there are eight internal sites for positive ions that have four O^{2-} ions as near neighbors (four-fold sites). There are also four internal sites for positive ions that have six O^{2-} ions as near neighbors (six-fold sites). In Fe_3O_4 the Fe^{2+} ions occupy the six-fold sites. Half of the Fe^{3+} ions occupy the six-fold sites and half occupy the four-fold sites. The electron spins in the six-fold sites are opposite to those in the four-fold sites so the Fe^{3+} ions cancel one another. The entire magnetism is due to the Fe^{2+} ions.

The term *ferrites* has been given to a series of compounds in which ions of several other elements substitute for Fe^{2+} ions. Table 22.2 lists several ions and the Bohr magnetons associated with each.

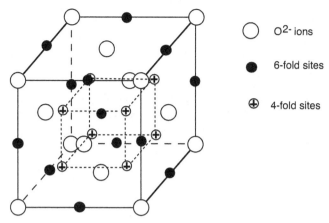

Figure 22.4. Structure of Fe_3O_4.

Table 22.2. *Bohr magnetons associated with several ions*

Ion	Bohr magnetons
Fe^{3+}	5
Mn^{2+}	5
Fe^{2+}	4
Co^{2+}	3
Ni^{2+}	2
Cu^{2+}	1
Zn^{2+}	0

Example Problem 22-2:

Calculate the saturation induction of $NiFe_2O_4$ which has a lattice parameter of 0.417 nm.

Solution: Magnetization is due solely to the Ni^{2+} ions in the four-fold sites.

$M = (2$ Bohr magnetons/$Ni^{2+})(9.27 \times 10^{-24}$ A$\cdot$m$^2)/(0.417 \times 10^{-9})^3 = 2.56 \times 10^5$ A/m.

$B_s = (4\pi \times 10^{-7}$ V$\cdot$s/A$\cdot$m$)(2.56 \times 10^5$ A/m$) = 0.32$ tesla.

The *B-H* Curve

In an alternating magnetic field, the resulting magnetization is described by the *B-H* curve (Figure 22.5). B_{max} is the largest maximum magnetic induction and B_r is the *residual magnetization* or *remanence* when the field is removed. H_c, called the coercive force, is the reverse field required to demagnetize the material. The area inside the loop is the *hysteresis* or energy loss per volume per cycle. Ideally a soft magnetic material has a small hysteresis, small H_c, and a high B_{max}.

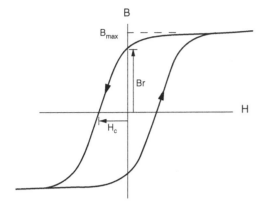

Figure 22.5. Typical *B-H* curve. From W. F. Hosford, *Physical Metallurgy*, CRC, 2005.

Soft Magnetic Materials

The ease of domain wall movement controls the hysteresis and coercive force. Movement of the boundaries between domains (*domain walls*) is hampered by grain boundaries and second phases. A soft magnetic material has a large grain size and few inclusions. If the grains are oriented with their directions of easy magnetization aligned with the applied fields, lower fields will be required to reach saturation, B_{max}. In transformers, eddy currents set up by the changing magnetic field lead to resistive losses so soft magnetic materials with high electrical resistivities are desirable. Iron containing 3% silicon is used in power transformers because of its high electrical resistivity, and because it is possible to align the grains with the easy direction of magnetization parallel to the direction of magnetization in the transformer. For extremely high frequencies, ceramic ferrites are used because of their extremely high resistivities.

Hard Magnetic Materials

Conversely, a high hysteresis and a high H_c are desirable in permanent magnets. The figure of merit for hard magnetic materials is $(B \times H)_{max}$. Figure 22.6 illustrates this. Obstacles to domain wall movement raise the maximum $B \times H$ product.

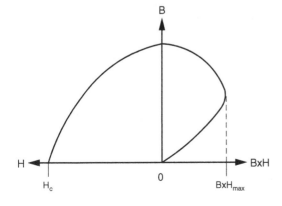

Figure 22.6. Second quadrant of the *B-H* curve (left) and a corresponding plot of the $B \times H$ product (right). From W. F. Hosford, *ibid.*

Desirable features are

1. small, isolated particles that are single domains;
2. elongated particles; and
3. a high magneto-crystalline energy.

The alloy alnico (8% Al, 15% Ni, 24% Co, 6% Cu, and 50% Fe) has these features (Figure 22.7).

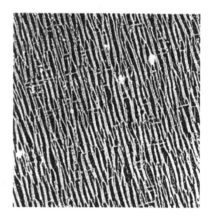

Figure 22.7. Microstucture of alnico V with elongated magnetic domains (dark) separated by non-magnetic regions (light). From R. M. Rose, L. A. Shepard, and J. Wulff, *Structure and Properties of Materials*, v. 4, Wiley, 1966.

Curie Temperature

There is a temperature above which a magnetic material ceases to be magnetic. Figure 22.8 shows the decrease of saturation magnetization with temperature.

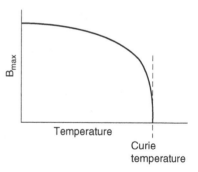

Figure 22.8. Decrease of saturation magnetization with temperature. The temperature at which B_{max} is zero is called the Curie temperature.

Magnetic Memory Storage

Magnetic materials for memory storage fall between hard and very soft magnetic materials. They are relatively soft but have square loops so that they are magnetized in either of two directions. Figure 22.9 shows a square loop.

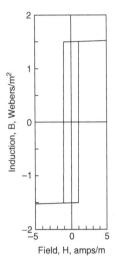

Figure 22.9. Decrease of saturation magnetization or memory storage.

Notes of Interest

Lodestone is magnetite, Fe_3O_4, which is ferromagnetic. It formed the basis for the first compasses. The Chinese recorded in the fourth century B.C. that iron was attracted to lodestone. The Chinese used compasses with lodestone as early as 1100 A.D. The compass was used by western Europeans in the late 1100s, by the Arabs in the early 1200s, and by the Vikings by 1300 A.D.

Problems

1. Figure 22.10 shows the second quadrant of the B-H curve for an alnico. Determine $B \times H_{max}$.

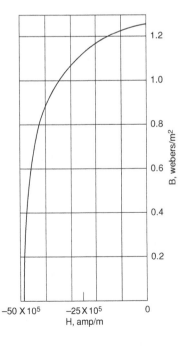

Figure 22.10. Second quadrant of the B-H curve for an alnico magnetic.

2. A steel containing 18% Cr, 8% Ni, and 0.08% C is non-magnetic after fast cooling from 900 °C. However, it becomes magnetic when it is deformed at –200 °C. Explain.

3. Consider the *B-H* curves in Figure 22.11.
   a. Which alloy has the highest coercive force?
   b. Which alloy has the highest residual magnetization (remanence)?
   c. Which alloy has the greatest hysteresis loss?
   d. Which alloy would make the best permanent magnet?
   e. Which alloy would be preferred for a memory device?
   f. In which alloy are the highest fractions of the domains aligned with the direction in which the field had been applied after the field is removed?
   g. Which alloy probably has the highest mechanical hardness?

4. Calculate the magnetization of nickel.

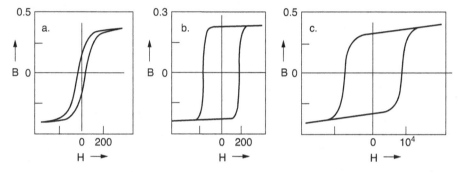

Figure 22.11. *B-H* curves for three magnetic alloys. The units of *B* are Webers/m^2 and the units of *H* are A/m.

23 | Corrosion

Corrosion of metals can be classified as either corrosion in aqueous solutions or as direct oxidation at high temperatures. Both are electrochemical in nature.

Aqueous Corrosion

An aqueous corrosion cell consists of an anode where metal ions go into solution and electrons are produced, a cathode where electrons are consumed, an aqueous solution between the anode and cathode, and an external electrical connection between the anode and cathode (Figure 23.1). The anode reaction can be written $M \rightarrow M^{n+} + ne^-$, where M stands for a metal. There are several possible cathode reactions:

$M^{n+} + ne^- \rightarrow M$ (This can occur only if there is a high concentration of M^{n+} ions in solution.)

$2H^+ + 2e^- \rightarrow H_2$ (This can occur only in an acid solution.)

$O_2 + 2H_2O + 4e^- \rightarrow 4(OH)^-$ (This is the most common cathode reaction. Note that it requires dissolved O_2.)

$O_2 + 4H^+ + 4e^- \rightarrow 2H_2O$ (This occurs in acidic solutions and there must be O_2 in an acid solution.)

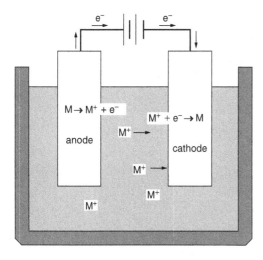

Figure 23.1. Corrosion cell. The anode is where electrons are generated in the external circuit, and the cathode is where they are consumed.

The anode and cathode reactions must occur at the same rate. The corrosion rate is often limited by the cathode reaction.

The material of the anode is more active (less noble) than the material of the cathode. The electromotive series (Table 23.1) lists the relative activity of common metals in one-molar solutions of their own salts. The most noble (least reactive) metals appear at the top and most reactive at the bottom. These potentials were determined in cells with semi-permeable membranes and each electrode in a one-molar solution of its salt as indicated in Figure 23.2.

Table 23.1. *Electode potentials (25°C; molar solutions)**

Anode half-cell reaction (the arrows are reversed for the cathode half-cell reaction)	Electrode potential used by electrochemists and corrosion engineers (V)	
$Au \rightarrow Au^{3+} + 3e^-$	+1.50	
$2H_2O \rightarrow O_2 + 4H^+ + 4e^-$	+1.23	
$Pt \rightarrow Pt^{4+} + 4e^-$	+1.20	
$Ag \rightarrow Ag^+ + e^-$	+0.80	
$Fe^{2+} \rightarrow Fe^{3+} + e^-$	+0.77	Cathodic (noble)
$4(OH)^- \rightarrow O_2 + 2H_2O + 4e^-$	+0.40	
$Cu \rightarrow Cu^{2+} + 2e^-$	+0.34	
$H_2 \rightarrow 2H^+ + 2e^-$	+0.000	
$Pb \rightarrow Pb^{2+} + 2e^-$	−0.13	Reference
$Sn \rightarrow Sn^{2+} + 2e^-$	−0.14	
$Ni \rightarrow Ni^{2+} + 2e^-$	−0.25	
$Fe \rightarrow Fe^{2+} + 2e^-$	−0.44	
$Cr \rightarrow Cr^{2+} + 2e^-$	−0.74	
$Zn \rightarrow Zn^{2+} + 2e^-$	−0.76	Anodic (active)
$Al \rightarrow Al^{3+} + 3e^-$	−1.66	
$Mg \rightarrow Mg^{2+} + 2e^-$	−2.36	
$Na \rightarrow Na^+ + e^-$	−2.71	
$K \rightarrow K^+ + e^-$	−2.92	
$Li \rightarrow Li^+ + e^-$	−2.96	

* From L. H. Van Vlack, *Elements of Materials Science and Engineering*, 3rd ed., Addison-Wesley, 1974, p. 414.

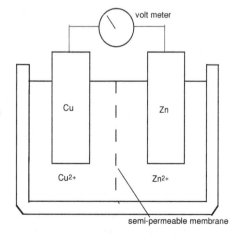

Figure 23.2. Corrosion cell for determining electrode potentials.

Example Problem 23-1:

The voltage between Cd^{2+} and Ag^{2+} in a cell similar to Figure 22.2 was found to be 1.2 V. Where should the reaction $Cd \rightarrow Cd^{2+} + 2e^-$ be in Table 23.1?

Solution: If the voltage were $+1.2$ V, cadmium would be more noble than gold, so the voltage must be -1.2 V, placing the reaction in the table at $0.8 - 1.2 = -0.4$ V.

The relative activity of a metal changes somewhat in different solutions and depends on the concentration of ions in solution. Table 23.2, called the galvanic series, lists various metals and alloys according to their electropotential in seawater.

Table 23.2. *Galvanic series of several metals and alloys*

Graphite	Least active
Silver	
18-8 Cr-Ni stainless steel (P)*	
Nickel (P)	
Copper	
Brass	
Tin	
Lead	
Nickel (A)	
18-8 Cr-Ni stainless steel (A)	
Plain carbon steel	
Aluminum	
Zinc	
Magnesium	Most active

* P signifies passive state; A signifies active state.

Differences in oxygen concentration from one region to another will create corrosion cells within a single material. The region where the oxygen concentration is high becomes the cathode, and where the oxygen concentration is low becomes the anode. Figure 23.3 illustrates several oxygen concentration cells.

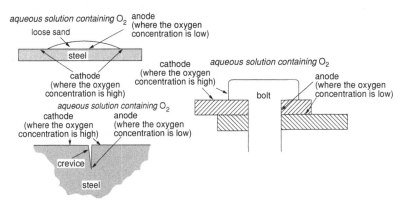

Figure 23.3. Oxygen concentration cells. The regions that are shielded from oxygen are the anodes, whereas the cathode reaction occurs where oxygen is plentiful.

Small differences in electropotential within a metal often arise from differences in composition, differences in the amount of cold work (Figure 23.4), or even from grain boundaries (Figure 23.5).

Figure 23.4. Regions that have been cold worked are anodic to regions that have not.

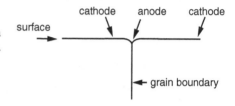

Figure 23.5. Because atoms at grain boundaries are in a higher energy state, the grain boundaries become anodic.

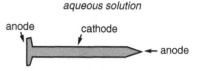

Passivity

Some highly oxidizable metals can become very inert to corrosion in an oxidizing solution. This condition is called *passivation*. A very thin adsorbed oxygen layer on stainless steels is sufficient to cause them to be passive. Oxygen and a very small amount of corrosion are required to maintain the passive state. Stainless steels containing 12% or more chromium are passive under most conditions if sufficient oxygen is present. Aluminum and titanium alloys may be passive under special conditions. They are passive only with sufficient oxygen and in the absence of chloride ions.

Corrosion Protection

Metals can be protected against corrosion by reversing the voltage in the corrosion cell. This can be done with an imposed DC voltage or by having contact with a more reactive metal. Steels plated with zinc (galvanized) are protected by the zinc (Figure 23.6).

Figure 23.6. (A) Plating steel with zinc (galvanizing) offers cathodic protection to steel if the plating is scratched. (B) Tin plating offers no cathodic protection to the steel if the plating is scratched.

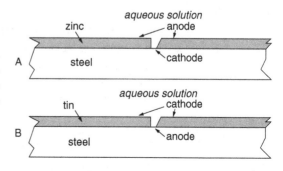

Zinc and magnesium are used to protect steel ships and buried pipes (Figure 23.7).

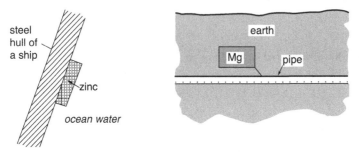

Figure 23.7. Corrosion protection by sacrificial corrosion of zinc (left) and magnesium (right).

Rust

Rust is ferric oxide, Fe_2O_3, or ferric hydroxide, $Fe(OH)_3$. Ferrous ions, Fe^{2+}, are soluble but further oxidation produces ferric ions: $3Fe^{2+} + 9OH^- \rightarrow 3Fe(OH)_3$. Ferric hydroxide is insoluble so it precipitates. If dried, ferric hydroxide turns to an oxide: $2Fe(OH)_3 \rightarrow Fe_2O_3 + 3H_2O$. Often the rust-producing reaction occurs at some distance from where the anode reaction occurs, so rust deposits may not be directly over the corroded region. This is illustrated in Figure 23.8.

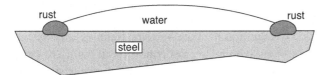

Figure 23.8. Rust formation away from corroded sites where the oxygen concentration is higher.

Direct Oxidation

Even direct oxidation in air at high temperature involves an electrolytic cell with an anode and a cathode. The anode reaction is $M \rightarrow M^{n+} + ne^-$ and the cathode reaction is $O_2 + 4e^- \rightarrow 2O^{2-}$. Either O^{2-} ions or M^{n+} ions and e^- must diffuse through the oxide. Because M^{n+} ions are smaller than O^{2-} the diffusion of M^{n+} ions controls the rate of oxidation. Figure 23.9 illustrates the reactions and transport in direct oxidation.

Al_2O_3 and Cr_2O_3 have very few defects so diffusion and electron transport is very slow. Hence, they are very protective. For an oxide to be protective, it must cover the surface. Hence, the volume of oxide is greater than or equal to the volume of metal oxidized. If $nM + mO \rightarrow MnOm$, protection occurs if

$$(MW)_{oxide}/\rho_{oxide} \geq n(AW)_M/\rho_M. \tag{23.1}$$

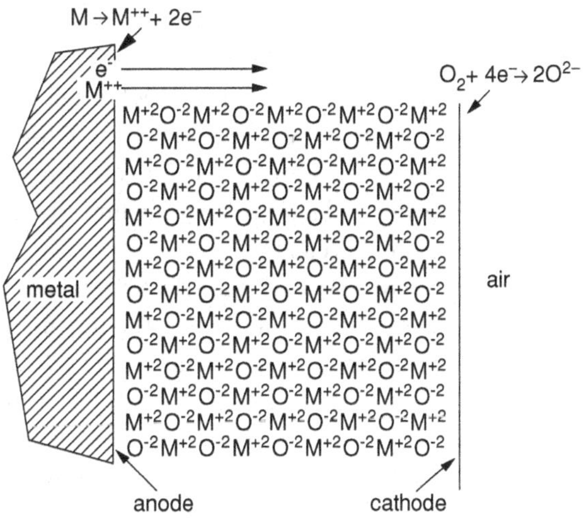

Figure 23.9. Direct oxidation. Oxide forms by diffusion of anions and electrons to the oxide-air surface.

Equation 23.1 is not satisfied for alkali or alkaline earth metals, so they oxidize rapidly in hot air. However, if the ratio $[(MW)_{oxide}/\rho_{oxide}]/[n\,(AW)_M\rho_M]$ is too high, the compressive stresses in the oxide may cause it to spall off. This is the case for Fe_2O_3. Another necessary condition for a protective oxide is that it must be solid. Tungsten and molybdenum oxidize very rapidly at high temperatures because their oxides are volatile. Because vanadium pentoxide (V_2O_5) forms a low-melting eutectic with Fe_2O_3, which flows off of the surface, fuels containing a small amount of vanadium have caused serious problems in power-generating turbines.

Notes of Interest

In about 1815 Francis Wollaston first suggested that acid corrosion was electrochemical. In 1824 Sir Humphrey Davy showed that there was a voltage between two different metals when they are connected and immersed in water. From this work he realized that there was a voltage set up between them and suggested that the copper bottoms of ships could be protected by attaching iron or zinc plates to them. Auguste-Arthur De la Rive, a Swiss scientist, showed in 1830 that zinc corroded when in contact with many metals. He arranged metals in order of activity and also showed that this order was dependent upon the electrolyte. The generally accepted theory assumed that an electrochemical reaction demanded the presence of two metals, or a metal and a metal oxide. In about 1830, Theodore Sturgeon showed that there could be potentials between different regions on the same metal surface. He found that differences of electrolyte concentration and temperature caused potential differences.

Thomas Edison tried to cathodically protect ship hulls with impressed voltages but his scheme failed.

In the 1930s Carl Wagner showed that high-temperature oxidation was also electrochemical in nature, with the oxide being the electrolyte, the metal acting as the anode, and the air-oxide surface acting as the cathode.

Problems

1. Steel is often plated with other metals. Which of the following metals would offer galvanic protection to steel in sea water?
   a. Ni, b. Zn, c. Al, d. Sn, e. Cu, f. Mg.

2. Copper is sometimes recovered from mine water by immersing steel scrap in the water and later recovering fine copper powder. Write the appropriate anode and cathode reactions.

3. Sometimes corrosion data are given in weight loss per area per time. For example, in the corrosion of steel, the loss might be 0.15 lbs/ft^2 per year. If uniform attack is assumed, what would be the corrosion in terms of mpy? The unit mpy means thousandths of an inch (mil) per year.

4. Which is worse from a corrosion standpoint, aluminum plates riveted together with steel rivets, or steel plates riveted together with aluminum rivets?

5. If steel is galvanized with 0.50 oz. of zinc per square foot, what is the thickness of the zinc plate?

6. Magnesium parts are heat treated in an atmosphere of 1% SO_2 which forms a coat of $MgSO_4$ on the surface. Why is this atmosphere preferable to air? The densities of $MgSO_4$, MgO, and Mg are 2.66, 3.85, and 1.74, respectively.

7. a. Tungsten has a melting point of 3400 °C. Why is it not considered for use in jet engines?
   b. What advantages do aluminum alloys have over more refractory materials at operating temperatures of 400 °F?

24 Modern Manufacturing Techniques, Surface Treatments, and Recycling

Two-dimensional Photolithography

The intricate circuits formed on semiconductor chips are patterned by *photolithography*. A photosensitive material (*photoresist*) is applied to a semiconductor surface and baked at a low temperature. The surface is then exposed to intense ultraviolet light. The exposed regions of the photoresist become soluble in a developer and are removed. The surface is then baked at a somewhat higher temperature to harden the remaining photoresist. The exposed regions are then acid etched.

The etched regions may then be treated differently from the rest of the surface. They may be *doped* by exposure to plasma containing n- or p-type impurities, or they may be plated to form conducting circuits. Integrated circuits are composed of many overlapping layers (Figure 24.1), each defined by photolithography. *Dopants* are diffused into the substrate, and additional ions are implanted into some layers. Metal and polycrystalline silicon are deposited on others to form conductive circuits.

Figure 24.1. Integrated circuit with many layers. From Wikipedia article on integrated circuits, 2007.

Transistors can be formed where there are n-p-n or p-n-p diffusion layers. Resistors are formed by meandering stripes of varying lengths. Capacitors are formed by parallel conductors separated by insulating material. The most common integrated circuits and those with the highest density are random access memories.

Photo-stereolithography

Three-dimensional parts can be made by photo-stereolithography using a liquid that polymerizes on exposure to ultraviolet light. A substrate platform is lowered into a bath and a desired portion of the surface of the liquid is exposed to a beam of ultraviolet light, causing it to polymerize. A computer data file controls the area to be exposed to the scanning beam. After the surface is scanned, the substrate is lowered allowing an additional thin layer of liquid to flow over the surface. The process is repeated until a three-dimensional structure is polymerized. Finally, the liquid that has not polymerized is drained off, leaving a solid part.

Dense sintered parts of inorganic materials can be formed by adding sinterable ceramics or metals to the liquid and firing the polymerized shape to drive off the organics. Typically the liquid may contain 40 to 70% sinterable ceramic and/or metallic particles, 10 to 35% photocurable monomer, and 1 to 10% photo-initiators and dispersants.

This process has been used to make scaffolds for bone growth that can be implanted where bones are missing, for internal cores to allow air passages in cast turbine blades, and for other parts that require intricate ceramic shapes. Allowance must be made for shrinkage during sintering.

Laser Beams

The advantage of lasers is their high power and good focusability. They can create a great amount of heat in a selected region. This heating can be used for hardening or annealing of metals, vaporization to remove material, or to melt and rapidly solidify to form new phases of alloys. Laser beams are used to make very tiny holes for dies, in welding, to cut thin discs from silicon single crystals, and in photolithography.

Laser drilling is especially adaptable for small holes with large depth-to-diameter ratios. With laser drilling, a wide range of hole diameters is obtainable. Material such as steel, nickel alloys, aluminum, copper, brass, borosilicate glass, quartz, ceramic, plastic, and rubber have all been successfully laser drilled. The laser is so fast and so repeatable that it is particularly ideal for high production volumes associated with fully automated or semi-automated tooling applications.

In laser cutting a high-power beam is focused onto the material to be cut. The material then melts, burns, or vaporizes away, leaving an edge with a high-quality surface finish. Industrial laser cutters are used to cut flat-sheet material as well as structural and piping materials.

A laser beam provides a concentrated heat source, allowing for narrow, deep welds and high welding rates. Laser beams have a high power density (on the order of 1 Mw/cm^2), so the heat-affected zones are very narrow. The depth of penetration is proportional to the amount of power supplied. The heating and cooling rates are high.

Electron Beams

Electron beams are finding increasing use in manufacturing. Their advantage over light in lithography is narrower beam width. Electron beams are used in welding to produce very narrow welds. The energy of the electrons is converted into heat, instantly vaporizing the metal under temperatures near 25,000 °C. The heat penetrates deeply, making it possible to weld much thicker work pieces than is possible with most other welding processes. However, because the electron beam is highly focused, the total heat input is actually much lower than that of other welding process. As a result, the effect of welding on the surrounding material is minimal, and like laser welding the heat-affected zone is small. Distortion is slight, and the work piece cools rapidly, and while normally an advantage, this can lead to cracking in hardenable steels.

Electron beams may also be used for local heat treatment or to harden whole surfaces. Because of their intense heat they can cause only a very thin surface layer to be heated.

Electrical Discharge Machining

Electrical discharge machining (EDM) removes material with electrical arcs between the work piece and an electrode. It can be used on very hard materials that would be difficult or impossible to machine by conventional means. The critical limitation is that the work piece be electrically conductive. The tool may be shaped, or may be a wire that has been programmed to move along the desired path very close to the work but does not touch the piece. Consecutive sparks produce a series of micro-craters on the work piece and remove material along the cutting path by melting or vaporization. The particles are washed away by a dielectric fluid.

EDM is widely used by the tool and die industry. It is used to make prototype and production parts, especially where production quantities are relatively small. It is used to make dies for coining jewelry and badges.

Plasma Coating

A plasma stream of argon or nitrogen, often with a small proportion, can reach temperatures of 5,000 to 25,000 °C. Typical power levels employed for high-velocity coatings are 40 to 80 kW DC. The additional energy available, coupled with slightly modified torch design, translates to both higher thermal energy and higher particle velocities. The resulting melting of the surfaces of even very refractory materials produces better coating integrity.

Ion Implantation

With high-energy beams of ions one material can be implanted into another, thereby changing the physical properties of the solid. Ion implantation is used in fabrication of semiconductor devices and in metal finishing. The ions introduce both a chemical change in the target, in that they can be a different element than the target, and a structural change.

The equipment for ion implantation typically consists of a source of the desired ions, an accelerator where the ions are accelerated by a field, and a target chamber where the ions impinge on the material to be implanted. The electric currents in ion implanting are typically small (microamperes). Therefore, only a small amount of material can be implanted at a time so ion implantation is used where the amount of chemical change required is small.

Typical ion energies are in the range 2 to 500 keV. Energies in the range 1 to 10 keV can be used but result in a penetration of only a few nanometers or less. Higher energies can also be used but cause great structural damage to the target.

Recycling of Metals

A large fraction (42%) of steel is from recycling. Steel and iron can easily be separated magnetically from other scrap. The scrap is either re-melted in an electric arc furnace or added to pig iron in a basic oxygen furnace. All grades of steel can be recycled because most alloying elements are oxidized during processing. Tin and copper are the exceptions, and there is concern in the steel industry about the gradual buildup of these *tramp* elements in steel.

There are two sources of scrap aluminum. *New* scrap is the offal of manufacturing processes resulting from trimming and occasional bad parts. Since the composition of new scrap is well known, it can be re-melted to make more of the same alloy. Other scrap (*old* scrap) comes from cans, engine blocks, cylinder heads, buildings, and other sources. Its composition is less well known. None of the alloying elements can be removed during melting so old scrap must be used to produce alloys with less critical compositions. Aluminum scrap is shredded and any lacquer is removed from cans before re-melting. Re-melting aluminum requires only 5% of the energy to produce virgin aluminum from bauxite and emits only 5% of the CO_2. The energy saving is 14 kWh/kg.

A substantial amount of aluminum is recycled. It is estimated that in 1998, 40% of the world's aluminum was produced by recycling. In the United States nearly half of aluminum beverage cans are recycled, the rest going into landfills. The recycle rate in Europe is also about 40%. Almost 90% of the aluminum cans are recycled in Norway, Sweden, and Switzerland. The rate in Brazil and Japan is also over 90%.

Recycled copper accounts for approximately 40% of the U.S. consumption of copper. Recycling uses about 15% of the energy to produce copper from the ore. In general the value of recycled copper is about 90% that of virgin copper. The primary sources are electrical wiring, plumbing pipes, and fixtures. Control of the scrap is important. To produce new wire, only high-conductivity (highly pure) scrap can be used. Some impurities may cause *hot shortness* (cracking during hot working). Copper and copper alloy scrap must be careful segregated. Pure copper, not contaminated by other metals, can be used to produce high-quality products. Alloy scrap that is segregated can be used for products of similar composition. Copper that is mixed with other metals, perhaps by having been tinned or soldered, or alloys to which lead has been added for machinability can be used in alloys that contain these metals. For example, many bronzes contain both lead and tin.

Brass extrusions and forgings are usually made by melting scrap of similar composition adjusted by the addition of virgin copper or zinc. If scrap has been contaminated beyond acceptable limits, it is necessary to re-refine it back to pure copper using conventional secondary metal refining techniques.

At present, approximately 70% of the zinc produced originates from mined ores and 30% from recycled or secondary zinc. The level of recycling is increasing in step with progress in zinc production technology and zinc recycling technology.

The supply of zinc-coated steel scrap increases annually as more auto-body steel is galvanized. About half of the world's steel is produced in electric arc furnaces. In the process flue dust with high zinc content is treated to recover zinc.

The U.S. Environmental Protection Agency (EPA) estimates that approximately 80% of all lead-acid car batteries are currently recycled. The lead battery plates are melted, refined, and recycled. The plastic case is shreded and recycled. Most battery recycling facilities will accept lead-acid batteries.

Different solder alloys contain different combinations of metal elements. The most common metals used in lead-free solder are tin, copper, silver, and sometimes zinc antimony and bismuth. Recycling these materials, the user can reduce the environmental impact and save reprocessing and disposal costs.

Photography (film, plates, and paper), electrical and electronics industry, jewelry, and spent catalysts are sources of silver for recycling. It is estimated that 26% of the annual usage of silver comes from recycling.

Recycling Plastics

Thermoplastics account for 80% of all plastics and these can be recycled. Heating of thermosetting polymers will cause them to decompose without melting, so they can only be used as landfill. The American Plastics Council has established a system of symbols (Figure 24.2) to identify several of the more common plastics for recycling:

1. Polyethyleneterephthalate (also indicated by PET or PETE and called "polyester"),
2. High-density polyethylene (also PE-HD or HDPE),

3. Polyvinylchloride (also PVC),
4. Low-density polyethylene (also PE-LD or LDPE for very-low-density polyethylene),
5. Polypropylene,
6. Polystyrene, and
7. Other resins or a combination of resins.

Figure 24.2. Recyling symbols for polymers.

Resins that have already been recycled are identified by an R before the previous designation. For example RHDPE means recycled high-density polyethylene.

Recycling polyethylene uses only 1/3 of the energy required to produce new polyethylene. According to the EPA, recyling of one pound of PET saves about 12.5 MJ. Generally recycled polymers are less desirable to manufactures than new resins and are used in less critical applications. For example, virgin PET is used for bottles, and recycled PET is used mainly for fibers for carpets and fiberfill. New HDPE is also used for bottles, but recycled HDPE is used for drainage pipes, laundry detergent bottles, floor tile, and picnic furniture. New PVC finds use as squeezable bottles, frozen food bags, clothing, and carpets while recycled PVC is used for floor tile, garbage can liners, furniture, trashcans, and lumber. Polypropylene (PP) is used in food and medicine packaging but recycled PP is used for brooms, battery cases, ice scrapers, trays, and pallets. Virgin polystyrene is used for rigid and foamed products including coffee cups. Among the recycled applications are foam packaging, thermal insulation, egg cartons, carryout containers, and rulers.

Recycling Glass and Paper

It is estimated that recycling of glass saves 50% of the energy to make new glass from sand and limestone and generates 20% less air pollution and 50% less water pollution. Recycling of paper uses only 40% of the energy to make paper from wood and causes only 5% as much air pollution.

Many products that claim to be biodegradable should more properly be called oxi-degradable. Plastarch material (PSMpolyactide) and polyactide (PLA) do degrade by composting. Products made from petrochemical compounds generally do not biodegrade. Tree leaves are biodegradable. They are made in the spring, used by the plant during the summer, drop to the ground in autumn, and are assimilated into the soil. There are no micro-organisms that can similarly break down most man-made products.

Notes of Interest

The development of the laser started in 1958 with the publication of a paper in *Physical Review, Infrared and Optical Masers* by Arthur L. Schawlow and Charles H. Townes. Schawlow and Townes had been interested in microwave spectroscopy, which had emerged as a powerful tool for puzzling out the characteristics of a wide variety of molecules in the 1940s and early 1950s. They planned to develop a device for studying molecular structures. Neither envisioned inventing a device that would revolutionize a number of industries, from communications to medicine.

Recycling has been practiced throughout human history. Even before the industrial revolution, scrap bronze, tin, and other precious metals were collected and melted down for reuse. The reason was economic: Recycled metals were cheaper than virgin metals. The shortage of materials during World War II greatly encouraged recycling. Governments in every country involved in the war urged citizens to donate metals and conserve fiber. Recycling programs established during the war were continued in some countries such as Japan after the war ended.

Problems

1. What is the advantage of the extremely narrow heat-affected zone during welding by laser or electron beams?
2. Compare the narrowness of beams obtainable by lasers and electrons.
3. Compare the advantages of induction hardening and carburizing for surface hardening.
4. It is estimated that it takes 15 kWh/kg to produce aluminum from bauxite. If it takes only 1/3 of this energy to produce aluminum from scrap, estimate what energy would be saved by recycling 1/2 of the 2 billion aluminum beverage cans consumed every year. A typical beverage can contains about 0.015 cm^3 of aluminum.
5. Examine the recycle codes on a number of plastic objects and cite objects with at least three different codes.

Wood

Structure

Wood is composed of hollow cells that transport sap in the living tree. Annual growth rings are evident in the cross section of a tree trunk as shown in Figure A1.1. They are easily seen because faster growth in the spring results in larger cells (Figure A1.2). Softwoods come from conifer (e.g., pine, spruce, and fir) and hardwoods from broadleaf trees (e.g., oak, maple, hickory, poplar, and willow). Most hardwoods are denser and harder than softwoods, but there are exceptions. Douglas fir is harder and denser than willow and poplar.

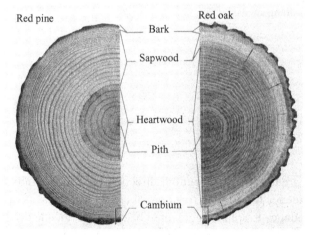

Figure A1.1. Cross sections of typical softwood and hardwood. From R. B. Hoadley, *Understanding Wood*, The Taunton Press, 1980.

Sapwood is the portion of the trunk through which sap is conducted. Heartwood is dead sapwood and is usually darker in color. The number of growth rings in the sapwood is between 20 and 40 for most hardwoods.

Bone-dry wood consists mainly of three compounds: cellulose (40 to 50%), hemicellulose (15 to 25%) in the cell walls, and lignin (15 to 30%), which holds the cells together. In a living tree, wood usually contains about 30% moisture. The shrinkage

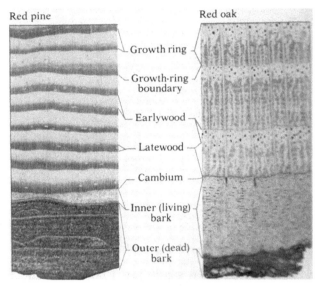

Figure A1.2. Growth rings in typical softwood and hardwood. From R. B. Hoadley, *ibid.*

on drying varies with direction. Table A1.1 lists the shrinkage of several woods on drying to 6% moisture.

Table A1.1. *Percent shrinkage on drying*[*]

Wood	Radial direction	Tangential direction
White pine	2.1	6.1
Douglas fir	4.8	7.6
Western cedar	2.2	4.9
Yellow birch	7.3	9.5
Hickory	4.9	8.9
Maple	3.7	7.1
White oak	5.6	10.5

[*] In all cases the longitudinal shrinkage is less than 0.25%.

Because the shrinkage is greater in the tangential direction than in the radial direction, there is a strong tendency for a log to form radial cracks when drying, as shown in Figure A1.3. This tendency to split is aggravated because the outside dries before the inside. To prevent this, lumber is usually cut from logs before oven drying to about 6% moisture.

Even after being sawn into board, the directional dependence of shrinkage causes warping as illustrated in Figure A1.4.

Dimensional Changes with Moisture

After drying, the moisture content of wood will approach equilibrium with the atmosphere. At 20 °C, the moisture content may vary from 4.5% if the humidity is 20%

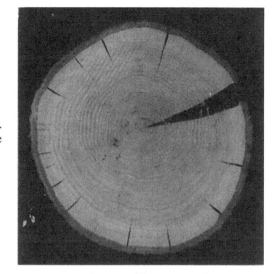

Figure A1.3. Splitting of a red oak log on drying. From R. B. Hoadley, *Understanding Wood*, The Taunton Press, 1980.

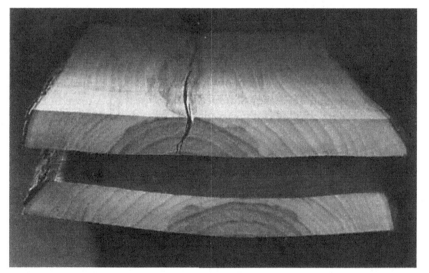

Figure A1.4. Elm board warped because the shrinkage in the tangential direction is greater than in the radial direction (bottom). Splitting when a warped board is flattened (top). From R. B. Hoadley, *ibid.*

to 16% if the humidity is 80%. The change of moisture content results in dimension changes that vary with direction as listed in Table A1.2.

Table A1.2. *Dimensional change with change of moisture content*

Radial direction	0.15% per % moisture change
Tangential direction	0.25% per % moisture change
Longitudinal direction	0.01% per % moisture change

Anisotropy of Properties

Table A1.3 shows that the elastic modulus and tensile and compressive strengths vary with direction.

Table A1.3. *Properties of common woods at 12% moisture content*

	White oak	Douglas fir
Density (Mg/m^3)	0.68	5.0
Young's modulus (GPa)	6.0	2.9
Tensile strength*		
Radial (MPa)	5.4	2.4
Compressive strengths (MPa)		
longitudinal	51	51
radial	7.4	5.2

* The longitudinal tensile strengths are about 20 times the radial tensile strengths.

Plywood

The orientation dependence of properties can be largely circumvented by the use of plywood, composed of plies with their grain oriented at 90° to one another. The plies are cut from rotating logs that have been softened with moisture as shown in Figure A1.5. The number of plies is always an odd number.

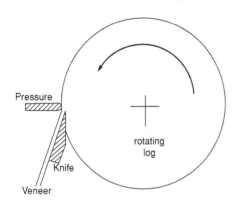

Figure A1.5. Cutting veneer for plywood.

Notes of Interest

The birth of the plywood industry began in 1907 when Portland Manufacturing installed an automatic glue spreader and a sectional hand press to produce plywood at a rate of 420 panels a day. Initially the market was primarily for door panels. Automobile running boards were made from plywood but this stopped because of water damage. In 1934 Dr. James Nevin, a chemist at Harbor Plywood Corporation in Aberdeen, Washington, finally developed a fully waterproof adhesive. This technology advancement had the potential to open up significant new markets.

After World War II, the industry produced 1.4 billion square feet of plywood. By 1954, the industry approached 4 billion square feet. By 1975 U.S. production alone exceeded 16 billion square feet.

Problems

1. Figure A1.6 shows corners of two picture frames. One was made in the winter and the other in the summer. Which frame was made in the winter?
2. A 250-mm-diameter pole initially contained 30% moisture. After a long time it developed a radial crack that was 25 mm wide at the surface. Estimate the moisture content at cracking.
3. Why do all sheets of plywood have an odd number of plies?
4. What compressive stress is developed in the circumferential direction of a fir board if it is prevented from expanding when its moisture content increases 5%?

Figure A1.6. Corners of two picture frames. One was made in the winter and the other in the summer. From R.B. Hoadley, *Understanding Wood*, The Taunton Press, 1980.

Miller Indices for Planes and Directions

It is often convenient to identify a plane or direction in a crystal by its indices. Note that all parallel planes have the same indices.

Planar Indices

The rules for determining the Miller indices of planes are as follows:

1. Write the intercepts of the plane on the three axes in order (a_1, a_2, and a_3).
2. Take the reciprocals of these.
3. Reduce to the lowest set of integers with the same ratios.
4. Enclose in parentheses (hkl).

Commas are not used except in the rare case that one of the integers is larger than one digit. (This is rare because we are normally interested only in planes with low indices.) If a plane is parallel to an axis, its intercept is taken as ∞ and its reciprocal as 0. If the plane contains one of the axes or the origin, either analyze a parallel plane or translate the axes before finding indices. This is permissible since all parallel planes have the same indices. A negative index is indicated by a bar over the index rather than a negative sign, e.g., ($11\bar{1}$). Figure A2.1 shows several examples.

Direction Indices

The indices of a direction are the translations parallel to the three axes that produce the direction under consideration. The rules for finding direction indices are as follows:

1. Write the components of the direction parallel to the three axes in order.
2. Reduce to the lowest set of integers with the same ratios.
3. Enclose in brackets [uvw].

Figure A2.2 shows several examples.

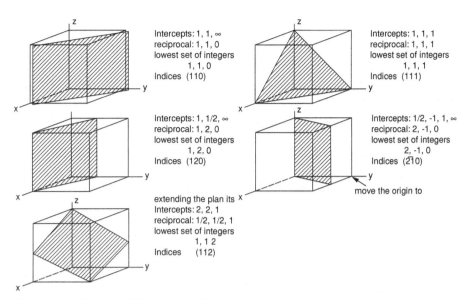

Figure A2.1. Finding Miller indices of several planes.

Families of Planes and Drections

A family of planes consists of those that are crystallographically equivalent. For example in a cubic crystal, (111), $(11\bar{1})$, $(1\bar{1}1)$, and $(1\bar{1}\bar{1})$ are equivalent. The family notation, $\{111\}$, means any or all of (111), $(11\bar{1})$, $(1\bar{1}1)$, and $(1\bar{1}\bar{1})$. Similarly a family of directions consists of those that are crystallographically equivalent. For example the family $\langle 210 \rangle$ consists of $[210]$, $[120]$, $[201]$, $[102]$, $[021]$, $[012]$, $[2\bar{1}0]$, $[1\bar{2}0]$, $[10\bar{2}]$, and $[01\bar{2}]$. Table A2.1 summarizes the types of brackets for specific planes, and directions and families of planes and directions.

Note of Interest

Welsh minerologist William Hallowes Miller first devised the system of Miller indices in his *Treatise on Crystallography* published in 1839.

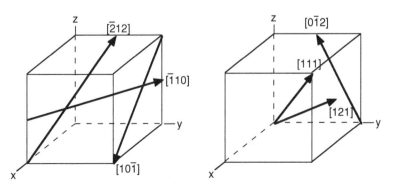

Figure A2.2. Examples of direction indices.

Table A2.1. *Types of brackets for planes and directions*

	Specific	Family
Planes	(hkl)	$\{hkl\}$
Directions	$[uvw]$	$\langle uvw \rangle$

Problems

1. Write the Miller indices for the planes sketched in Figure A2.3.
2. Write the direction indices for the directions sketched in Figure A2.4.
3. In a cubic crystal, how many planes are included in the $\{124\}$ family, not including opposite sides of the same plane (i.e., (124) is the same as $(\overline{124})$)?

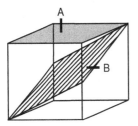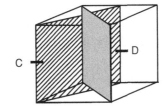

Figure A2.3. Several planes.

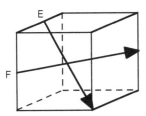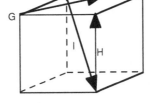

Figure A2.4. Several directions.

X-ray Diffraction

X-rays are used to determine crystal structures, lattice parameters of crystals, and orientation of crystals. Bragg's law describes the conditions for an X-ray beam to be defracted by a crystal:

$$n\lambda = 2d\sin\theta, \qquad\qquad \text{A3.1}$$

where λ is the wavelength of the X-rays, d is the spacing between lattice planes, θ is the angle of the incident beam to the diffracting planes, and n is an integer. This is illustrated in Figure A3.1. The upper and lower rays will be in phase if the extra distance that the lower ray travels, $2d\sin\theta$, equals an integral number of wavelengths.

By measuring the angle, θ, at which diffraction occurs, the distance between planes can be found from equation A3.1. The crystal structure and lattice parameter can be determined by measuring the angles of several diffraction peaks, identifying the corresponding planes, and calculating the spacing of the planes.

Note of Interest

William Lawrence Bragg was born in Australia. After graduating from the University of Adelaide, he continued his studies at Cambridge. He and his father were jointly awarded the Nobel Prize in Physics in 1915 for their publication of *X Rays and Crystal Structures*.

Figure A3.1. Bragg conditions for diffraction.

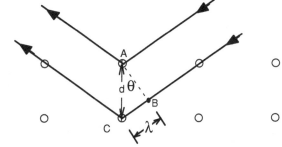

Problem

1. For an aluminum alloy irradiated with X-rays of wavelength 0.1537 nm, a diffraction peak occurs for reflection from {111} planes at an angle of 41.8°. Determine the lattice parameter.

Surfaces

There is an energy per area associated with every surface because molecules or atoms at a surface have different surroundings than those in the interior. The units of surface energy are J/m^2. The work to create a surface can be thought of as a surface tension on a line, working through a distance so surface tension in N/m is equivalent to surface energy.

Relation of Surface Energy to Bonding

An approximate calculation of surface energy can be made by envisioning a surface being formed by mechanical forces across them and calculating the work required to separate the two halves of a crystal. The surface energy, γ_s, is given by

$$\gamma_s = (1/2)(\text{work/area}) = (1/2) \int \sigma \, ds, \qquad \text{A4.1}$$

where σ is the stress to separate the surfaces and s is the separation distance.

Because both surface energies and melting temperatures are related to bonding strength, the surface energy is closely related to the melting point as shown in Figure A4.1.

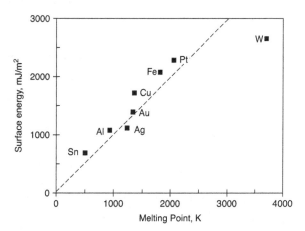

Figure A4.1. Correlation of surface energy with melting point. From W. F. Hosford, *Physical Metallurgy*, CRC, 2005.

Grain Boundaries

The energy of a grain boundary depends on the misorientation across the boundary. Low-angle tilt-grain boundaries are composed of edge dislocations (Figure A4.2). The angle of misorientation, θ, is given by

$$\theta = b/L, \qquad\qquad\qquad \text{A4.2}$$

where b is the Burgers vector and L is the distance between dislocations. The number of dislocations per length, n, is equal to the reciprocal of L:

$$n = 1/L = b/\theta. \qquad\qquad\qquad \text{A4.3}$$

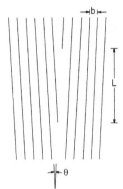

Figure A4.2. A low-angle tilt boundary is composed of edge dislocations.

The energy of an edge dislocation is given by $(Gb^2/4\pi)\ln(r_1/r_o)/(1+\upsilon)$, where G is the shear modulus, υ is Poisson's ratio, and r_o is a constant equal to about $b/4$. The value of r_1 is approximately equal to the distance between dislocations, so $r_1 = b/\theta$. Substituting the energy of a low-angle tilt boundary is $\gamma = n(Gb^2/4\pi)\ln(r_1/r_o)/(1+\upsilon)$ or

$$\gamma = (\theta/b)(Gb^2/4\pi)\ln[(4/\theta)]/(1+\upsilon). \qquad\qquad \text{A4.4}$$

At low angles, γ is proportional to θ, but at higher angles, γ/θ decreases as illustrated in Figure A4.3. Screw dislocations on a plane form twist boundaries. The misorientation across a low-angle twist boundary and its energy are proportional to the number of dislocations.

Segregation to Surfaces

Segregation of solutes to a grain boundary lowers its energy. Solutes that have a larger atomic size than the parent atoms occupy positions that are open, and smaller solute atoms occupy sites where there is crowding as shown schematically in Figure A4.4.

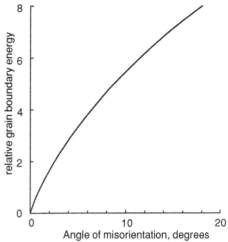

Figure A4.3. Relative energy of a low-angle tilt boundary calculated from equation A4.4.

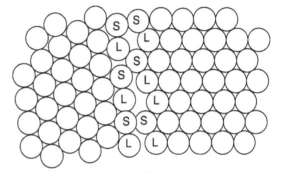

Figure A4.4. Two-dimensional sketch of a grain boundary. The crowding at atom positions indicated by S can be relieved if they are occupied by small atoms. Open positions, indicated by L, attract large atoms.

The ratio of grain boundary concentration to overall concentration, c_{gb}/c_0, decreases with increasing solubility. Hondras and Seah* showed that

$$c_{gb}/c_0 = A/c_{max},\qquad\qquad A4.5$$

where c_{max} is the solubility limit in the matrix and A is a constant with a value of about 1.

There is also segregation of solutes to free surfaces. One example is soapy water. Soap segregates to the surface, lowering the surface energy (surface tension).

Measurements of Relative Surface Energies

Most surface energies have been determined from other surface energies by measuring the angles at which surfaces meet. The angles at which three surfaces meet depend on the relative energies of the three interfaces (Figure A4.5). Each surface exerts a

* E. D. Hondras and M. P. Seah, "International Metallurgical Reviews," 1977, Review No. 222.

force per length on the junction that is equal to its surface tension. At equilibrium, the force vectors must form a triangle, so the law of sines gives

$$\gamma_{23}/\sin\theta_1 = \gamma_{31}/\sin\theta_2 = \gamma_{12}/\sin\theta_3. \qquad \text{A4.6}$$

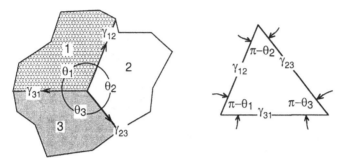

Figure A4.5. Relative surface energies. From W. F. Hosford, *Physical Metallurgy*, CRC, 2005.

Often two of the angles and two of the surface energies are equal. For example, consider the intersection of a grain boundary with a free surface. If the temperature is high enough and the time is long enough, the surface will thermally etch (by vaporization or surface diffusion) until an equilibrium angle is formed. From a balance of forces parallel to the grain boundary (Figure A4.6),

$$\gamma_{gb} = 2\gamma_{sv}\cos(\theta/2). \qquad \text{A4.7}$$

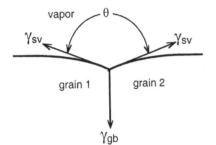

Figure A4.6. Intersection of a grain boundary with a free surface.

Surfaces of Amorphous Materials

A simple bond-breaking approach is not applicable to amorphous materials such as thermoplastics and glasses because they will adjust their molecular configuration to minimize the number of missing bonds. If there is more than one type of bond, the missing bonds at the surface are likely to be the weakest bonds and not characteristic of the overall bond strength.

Note of Interest

The reason that soapy water forms relatively stable bubbles is that in a bubble the surface tension must vary with location, being highest at the bottom and lowest at the top to balance gravitational effects. It is because the surface tension depends on the soap concentration at the surface and that this concentration is variable that stable bubbles can be formed.

Problems

1. Calculate the pressure inside of a 1-μm-diameter solid drop of gold, knowing that the surface energy of liquid gold is 1,485 mJ/m^2.

2. Find the equilibrium angle, θ, of the groove at a grain boundary in copper, knowing that for copper $\gamma_{gb} = 625$ mJ/m^2 and $\gamma_{sv} = 1,725$ mJ/m^2.

Dislocations

Edge and Screw Dislocations

Plastic deformation of crystalline solids occurs primarily by slip, which is the sliding of atomic planes over one another. An entire plane does not slide over another at one time. Rather it occurs by the motion of imperfections called *dislocations*. Dislocations are line imperfections in a crystal. The lattice around a dislocation is distorted so the atoms are displaced from their normal lattice sites. The lattice distortion is greatest near a dislocation and decreases with distance from it. One special form of a dislocation is an *edge* dislocation that is sketched in Figure A5.1. The geometry of an edge dislocation can be visualized by cutting into a perfect crystal and then inserting an extra half-plane of atoms. The dislocation is the bottom edge of this extra half-plane. An alternative way of visualizing dislocations is illustrated in Figure A5.2.

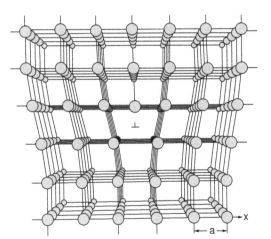

Figure A5.1. An edge dislocation is the edge of an extra half-plane of atoms. From A. G. Guy, *Elements of Physical Metallurgy*, Addison-Wesley, 1951.

An edge dislocation is created by shearing the top half of the crystal by one atomic distance perpendicular to the end of the cut (Figure A5.3B). This produces an extra half-plane of atoms, the edge of which is the center of the dislocation. The

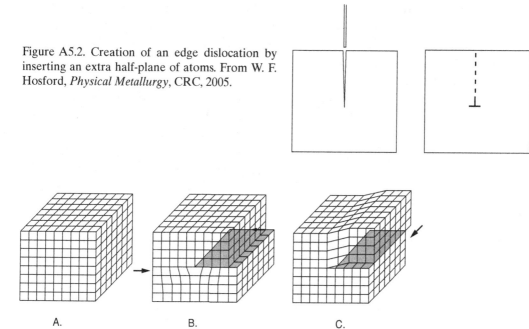

Figure A5.2. Creation of an edge dislocation by inserting an extra half-plane of atoms. From W. F. Hosford, *Physical Metallurgy*, CRC, 2005.

A.   B.   C.

Figure A5.3. Consider a cut made in a perfect crystal (A). If one half is sheared by one atom distance perpendicular to the end of the cut, an edge dislocation results (B). If one half is sheared by one atom distance parallel to the end of the cut, a screw dislocation results (C). From W. F. Hosford, *Mechanical Behavior of Materials*, Cambridge, 2005.

other extreme form of a dislocation is the *screw dislocation*. This can be visualized by cutting into a perfect crystal and then shearing half of it by one atomic distance in a direction parallel to the end of the cut (Figure A5.3C). The end of the cut is the dislocation. Around it, the planes are connected in a manner similar to the levels of a spiral ramp of a parking garage.

In both cases the dislocation is a boundary between regions that have slipped relative to each other. The direction of slip that occurs when an edge dislocation moves is perpendicular to the dislocation. In contrast, movement of a screw dislocation causes slip in the direction parallel to itself. The edge and screw are extreme cases. A dislocation may be neither parallel nor perpendicular to the slip direction.

The energy per length, U_L, of a screw dislocation is given by

$$U_L \approx Gb^2, \qquad\qquad\qquad \text{A5.1}$$

where G is the shear modulus and b is the Burgers vector, which is the shear displacement. For an edge dislocation the energy per length is somewhat higher,

$$U_L \approx Gb^2/(1 - v), \qquad\qquad\qquad \text{A5.2}$$

where v is Poisson's ratio.

Stress Fields

There is a hydrostatic stress, σ_H, around an edge dislocation. Figure A5.4 shows how the hydrostatic stress varies near an edge dislocation. The stress is compressive above the edge dislocation tensile below it. The dilatation causes interactions between edge dislocations and solute atoms.

In substitutional solutions, solute atoms that are larger than the solvent atoms are attracted to the region just below the edge dislocation where their larger size helps relieve the hydrostatic tension. Similarly, substitutional solute atoms that are smaller than the solvent atoms are attracted to the region just above the edge. In either case edge dislocation will attract solute atoms. In interstitial solid solutions, all solute atoms are attracted to the region just below the edge dislocation where they help relieve the tension. It is this attraction of edge dislocations in iron for carbon and nitrogen that is responsible for the yield-point effect and strain-aging phenomenon in low-carbon steel.

Edge dislocations of like signs tend to form walls with one dislocation directly over another as shown in Figure A5.5. The hydrostatic tension caused by one dislocation is partially annihilated by the hydrostatic compression of its neighbor. This relatively low-energy and therefore stable configuration forms a low-angle grain boundary.

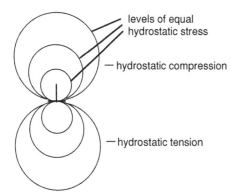

levels of equal
hydrostatic stress

— hydrostatic compression

— hydrostatic tension

Figure A5.4. Contours of hydrostatic stress around an edge dislocation. Note that the level of hydrostatic stress increases near the dislocation.

Figure A5.5. Low-angle tilt boundary formed by edge dislocations arranged so that the compressive stress field is partially annihilated by the tensile stress field of the one above.

Note of Interest

Dislocations were first postulated independently by M. Polyani, E. Orowan, and G. I. Taylor in 1934. It wasn't until the development of the transmission electron microscope decades later that dislocations were directly observed.

Problems

1. a. Calculate the energy per volume associated with dislocations in aluminum if there are 10^{15} m dislocations/m^3.

    b. If all of this energy were converted to heat, by how much would the temperature rise?

       Data: For aluminum, $G = 25$ GPa, $b = 0.286$ nm, $C = 0.94$ J/g °C, $\rho = 2.7$ Mg/m^3.

2. What is the distance between edge dislocations in a $1°$ tilt boundary?

APPENDIX 6

Avrami Kinetics

Avrami Kinetics

The rates of many solid-state reactions can be described by

$$f = 1 - \exp(-kt^n),$$ A6.1

where f is the fraction transformed, t is time, and k is a temperature-dependent constant. The exponent n is 4 or less. The effect of the exponent, n, on the transformation is shown in Figure A6.1. The values of k were not adjusted, so the apparent faster transformation rate with lower exponents is not real. Find an expression for the ratio of time for the reaction to be 90% complete to the time for it to be 10% complete and evaluate it for $n = 4$ and $n = 2$.

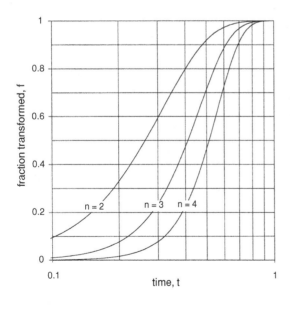

Figure A6.1. The effect of the exponent in the Avrami equation on the transformation. On a $\log(f)$–$\log(t)$ plot, the slope increases with n. From W. F. Hosford, *Materials Science: An Intermediate Text*, Cambridge, 2006.

Example Problem 1:

Find the ratio of the times for a reaction to be 90% complete to the time for it to be 10% complete. Compare that ratio for $n = 4$ and $n = 2$.

Solution: From equation 1, $\ln(1 - f) = -kt^n$. Comparing two degrees of completion, $\ln(1 - f_2)/\ln(1 - f_1) = (t_2/t_1)^n$ so $t_2/t_1 = [\ln(1 - f_2)/\ln(1 - f_1)]^{1/n}$.

Substituting $f_2 = 0.9$ and $f_1 = 0.1$, $t_2/t_1 = 21.8^{1/n}$. For $n = 4$, $t_2/t_1 = 2.16$, for $n = 2$, $t_2/t_1 = 4.7$.

The exponent n depends on nucleation and growth. If the nucleation rate is constant and growth occurs in three dimensions at a constant rate, $n = 4$. However, there are several reasons why the Avrami exponent may be less than 4. The nucleation rate may decrease with time because most favorable nucleation sites are used up early. In the extreme, it is possible that all nucleation sites are used up at the very start so nucleation makes no contribution to the exponent. The growth rate may decrease with time. This is true for precipitation from solid solution. For precipitation, the rate of growth is inversely proportional to the square root of time. Finally, if growth is in only one or two dimensions instead of three, growth will contribute less to the exponent.

The constant k, which depends on both the nucleation and growth rates, is very temperature sensitive. Changes in k shift the curve horizontally on a semi-logarithmic plot but do not change its shape. Figure A6.2 shows the rate of transformation for two temperatures, T_1 and T_2, with $n = 4$.

For 50% transformation, $-kt_{(0.5)}n = \ln(0.5)$, where $t_{(0.5)}$ is the time for $f = 50\%$ so

$$k = .69/t_{(0.5)}{}^n.$$ A6.2

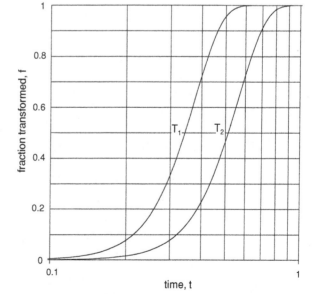

Figure A6.2. The effect of temperature on transformation kinetics. T_1 is higher than T_2. From W. F. Hosford, *Materials Science: An Intermediate Text*, Cambridge, 2006.

fraction transformed, f

time, t

Example Problem 2:

A reaction $\dot{N}$ is 20% complete after 45 s and 85% complete after 1.25 min. Determine the value of n in the Avrami equation.

Solution: Writing the Avrami equation as $-\ln(1-f) = bt^n$ and evaluating at two conditions, $-\ln(1-f_2)/-\ln(1-f_1) = bt_2^n/bt_1^n = (t_2/t_1)n$ so $n = \ln[\ln(1-f_2)/\ln(1-f_1)]/\ln (t_2/t_1) = \ln(\ln0.8/\ln0.15)/\ln(45/1.25 \times 60) = 4.2$. Many reactions, including recrystallization, can be described by Avrami kinetics.

Note of Interest

Equation A6.1 was first developed with an exponent of 4 by A. N. Kolmogorov in 1937 (*Izv. Akad. Nauk SSSR. Ser. Mat.*, 3, 1937, p. 355) and by W. A. Johnson and R. F. Mehl in 1939 (*Trans. AIME*, 135, 1939, p. 416) to explain the kinetics of pearlite formation in steel. In 1939 M. Avrami (*J. Chem. Phys.*, 1939, p. 1103) showed that the equation could be applied to other transformation reactions with different exponents.

Problems

1. Determine the exponent n for recrystallization of copper at 113 °C using the data in Figure 6.3.
2. Figure A6.3 gives data on a phase transformation.

   a. Determine the exponent in the Avrami equation.
   b. At what time would you expect the fraction transformed to be 0.001?
   c. At what time would you expect the fraction transformed to be 0.999?

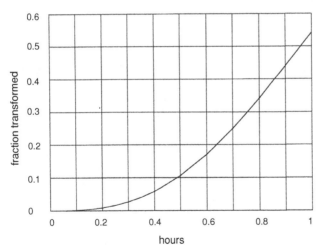

Figure A6.3. A plot for a phase transformation of the fraction transformed as a function of time. From W. F. Hosford, *ibid*.

Organic Chemistry

Bonds

Each element in an organic compound has a fixed number of bonds, each bond consisting of a pair of shared electrons. For example, the carbon atom has four bonds. Occasionally two carbon atoms share four electrons. In this case it has two bonds (or a double bond). Table A7.1 lists the number of bonds characteristic of various atoms.

Table A7.1. *Bonds associated with several elements*

Elements	Number of bonds
H, Cl, F	1
O, S	2
N	3
C, Si	4

Bond Energies

Each bond has a characteristic energy and bond length. These are given in Table A7.2.

Example Problem 1:

What is the longest wavelength of light that can break a carbon-carbon bond?

Solution: $\lambda = c/\upsilon = ch/E$. Substituting $c = 2.99 \times 10^8$ m/s, $h = 6.62 \times 10^{-24}$ Js, and $E = (370 \times 10^3 \, \text{J/mole})(6.02 \times 10^{23}/\text{mole})$, $\lambda = 323$ nm, which is in the ultraviolet.

Example Problem 2:

How much heat is released when a gram of ethylene polymerizes? The reaction is $n(CH_2 = CH_2) \rightarrow -\{CH_2–CH_2\}_n$. Knowing that the specific heat of polyethylene is

Table A7.2. *Bond energies and lengths*

Bond	Energy (kJ/mole)*	Length (nm)
C–C	370	0.154
C=C	680	0.13
C≡C	890	0.12
C–H	425	0.11
C–N	305	0.15
C–O	360	0.14
C=O	535	0.12
C–F	450	0.14
C–Cl	340	0.18
O–H	500	0.10
O–O	220	0.15
O–Si	375	0.16
N–H	430	0.10
N–O	250	0.12
F–F	160	0.14
H–H	435	0.074

* Energies required to break bonds and heat released when
they are broken.

1850 J/kg °C, what would the temperature rise be if the process were adiabatic (with no heat transferred to the surroundings)?

Solution:

From Table A7.1, the heat released is (370 kJ/mole)/(18 g/mole) = 20 kJ/g $\Delta T = (20\,\text{kJ/g})/(1.85\,\text{J/g}\,°C) = 10,800\,°C$. Obviously the material needs to be cooled during the reaction.

Bond Angles

When carbon and silicon atoms are bonded to other atoms of the same species (e.g., diamond, silicon, methane, carbon tetrachloride), the bonds are at tetrahedral angles (109.5°). When the surrounding atoms are not all the same, there are other characteristic angles. Some of these are listed in Table A7.3. A number of simple molecules involved in polymerization reactions are listed in Table A7.4.

Table A7.3. *Characteristic bond angles (degrees)*

Diamond	C	C–C–C	109.5
Ammonia	NH_3	H–N–H	107
Water	H_2O	H–O–H	104
Hydrogen sulfide	H_2S	H–S–H	92

Note of Interest

Until 1828 a theory known as *Vitalism* proposed that living organisms were necessary to make organic compounds. In 1828, German chemist Friedrich Wöhler (1800–1882) synthesized urea, H_2NCONH_2, which is an organic substance found in the urine of many animals, from the inorganic compound ammonium cyanate, NH_4OCN.

Table A7.4. *Molecules involved in polymerization reactions.*

Family	Characteristic Bond	Compound		
Alcohols	C-OH	H H-C-OH H methanol	H H H-C-C-OH H H ethanol	H H H H-C-C-C-OH H H H propyl alcohol
Aldehydes & keytones	C=O	H H C=O formaldehyde	H H H C-C=O acetaldehyde	R-C- R′ O keytone
Acids	C OH / O	HC OH / O formic acid	H HC-C OH / O H acetic acid	
Ethers	C-O-C	H H H C-O-C H H H dimethyl ether		
Esthers	C-O-C / O	H H H HC-C-O-C- C H ethylacetate H O H H		
Amides	C-N / H O H	H H HC-C-N H O H acetamide	O=C N H H / N H H urea (diamide)	
Amines	C-N H / H	H H H N-(CH2)6 -N H hexamethyldiamide		
Benzene (aromatics)	C—C / C / C=C	H H C—C HC / CH C=C H H benzene	H OH C—C HC / CH C=C H H phenol	

Problems

1. Why does the success of commercial chemical processes involving synthesis depend on heat removal?
2. How many C=C bonds are eliminated per mole of ethylene during the synthesis of polyethylene? How many C−C bonds are formed? How much heat is released per kilogram of polyethylene?

Average Molecular Weight

Molecular Weights

The molecular weight of a single molecule is simply the sum of the atomic weights of the atoms that it contains. Polymers generally contain molecules of different lengths. There are two ways of determining an average molecular weight from a distribution of molecular weights if the molecular weights are grouped. One, called the *weight-average molecular weight*, is the sum of weights of each group times the molecular weight of that group divided by the total weight:

$$\text{Wt. Av. MW} = \sum[(\text{MW of a group})(\text{weight of the group})]/(\text{total weight}). \quad \text{A8.1}$$

The second, called the *number-average molecular weight*, is the number of molecules (or moles) in each group times the molecular weight of that group divided by the total number of molecules (or moles):

$$\text{No. Av. MW} = \sum[(\text{MW of a group})(\text{mole of that group})]/(\text{total number of moles}). \quad \text{A8.2}$$

Example Problem 1:

Calculate the weight-average molecular weight and number-average molecular weight of a polymer with the following molecular weight distribution:

Molecular weight group (g/mole)	Weight of group (g)	Weight of moles
10,000	2000	0.2
12,000	4500	0.375
14,000	5000	0.357
16,000	3000	0.1875
18,000	1500	0.0833
total	16,000	1.203

Solution: The weight-average molecular weight $=$ [(10,000)(2000) $+$ (12,000) (4500) $+$ (14,000)(5000) $+$ (16,000)(3000) $+$ (18,000)(1500)]/16000 $=$ 13,690.

The number-average molecular weight $=$ [(10,000)(0.2) $+$ (12,000)(0.375) $+$ (14,000)(0.357) $+$ (16,000)(0.1875) $+$ (18,000)(0.0833)]/1.203 $=$ 13,300.

Note of Interest

Avogadro's number, 6.02214×10^{23}, is named after the Italian chemist Amedeo Avogadro, who in the early nineteenth century proposed that the volume of an ideal gas is proportional to the number of molecules, but never attempted to measure it. The first measurement of Avogadro's number was by an Austrian, Johann Josef Loschmidt, in 1865.

Problems

1. Calculate the weight-average molecular weight of a polymer that has 0.5 moles of MW $=$ 10,000, 1.0 moles of MW $=$ 15,000, and 1.5 moles of MW $=$ 20,000.
2. Calculate the number-average molecular weight of a polymer composed of 50 grams of MW $=$ 10,000, 100 grams of MW $=$ 15,000, and 100 grams of MW $=$ 20,000.

Bond Geometry in Compounds

Coordination

The structure of ionic crystals usually corresponds to the maximum possible coordination number. If the ions are assumed to be hard spheres, there must be contact between ions of opposite signs and no contact between ions of like sign. The coordination depends on the anion-to-cation diameters ratio.

Figure A9.1 shows that the critical condition for three-fold coordination is $R/(R + r) = \cos 30° = \sqrt{3}/2$. Therefore,

$$r/R \geq 0.1547. \tag{A9.1}$$

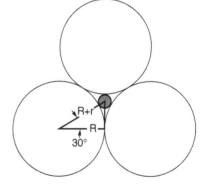

Figure A9.1. For three-fold coordination $r/R \geq 0.155$. From W. F. Hosford, *Materials Science: An Intermediate Text*, Cambridge, 2006.

Four-fold or tetrahedral coordination corresponds to the smaller ion at the center of a tetrahedron (see Figure A9.2). Conditions for four-fold coordination can be easily analyzed by imagining the tetrahedron inside a cube with edges of length a. Then $[2(R+r)]^2 = 3a^2$. At the critical condition, $(2R)^2 = 2a^2$, so $4(R + r)^2 = 6R^2$,

$$r/R \geq \sqrt{3/2}) - 1 = 0.2247. \tag{A9.2}$$

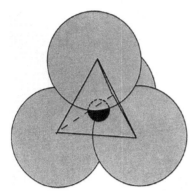
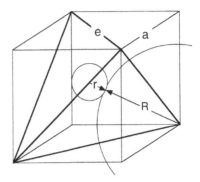

Figure A9.2. For four-fold tetrahedral coordination, $r/R \geq 0.2247$. From W. F. Hosford, *ibid.*

Figure A9.3 shows that six-fold or octahedral coordination corresponds to the smaller ion at the center of an octahedron with the larger ions on the corners. The critical ratio corresponds to $[2(r+R)]^2 = 2R^2$, so

$$r/R \geq \sqrt{2} - 1 = 0.4142. \hspace{2cm} \text{A9.3}$$

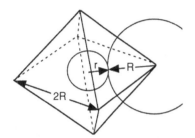

Figure A9.3. For six-fold (octahedral) coordination, $r/R \geq 0.4142$. From W. F. Hosford, *ibid.*

With eight-fold or cubic coordination (Figure A9.4), the smaller ion is in the center of a cube with larger ions on the corners. $[2(r+R)]^2 = 3(2R)^2$, so

$$r/R \geq \sqrt{3} - 1 = 0.7321. \hspace{2cm} \text{A9.4}$$

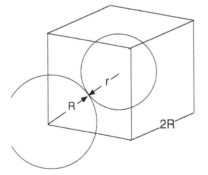

Figure A9.4. For eight-fold (cubic) coordination, $r/R \geq 0.7321$. From W. F. Hosford, *ibid.*

These geometric restrictions are summarized in Table A9.1.

Table A9.1. *Geometric restrictions on coordination*

Coordination	Minimum r/R ratio
Three-fold	0.1547
Four-fold	0.2247
Six-fold	0.4142
Eight- fold	0.7321

Ionic Radii

Table A9.2 gives the ionic radii determined by Pauling.

Structure with four-fold coordination includes both the zinc blende and wurtzite structures. Compounds with the zinc blende structure include αZnS, ZnO, SiC, BeO, AlP, GaP, αCdS, HgS, βAgI, InP, BeSe, AlAs, GaAs, CdSe, HgSe, CuI, InSb, BeTe, and AlSb.

Compounds with the wurtzite structure include βZnS, ZnO, SiC, MgTe, and CdSe.

Structure having six-fold coordination include the halite structure. Some of these compounds are MgO, CaO, SrO, BaO, CdO, MnO, FeO, CoO, NiO, KCl, KI, and KBr.

The hexagonal nickel arsenide structure also has six-fold coordination, including NiAs, FeS, FeSe, and CoSe.

Very few compounds have eight-fold coordination. These include CsCl and CsBr.

Table A9.2. *Ionic radii*[*]

Ion	Radius (nm)	Ion	Radius (nm)
Li^+	0.068	Be^{2+}	0.035
O^{2-}	0.140	F^-	0.133
Na^+	0.097	Mg^{2+}	0.066
Al^{3+}	0.051	Si^{4+}	0.042
S^{2-}	0.184	Cl^-	0.181
K^+	0.133	Ca^{2+}	0.099
Ti^{4+}	0.068	Cr^{3+}	0.063
Mn^{2+}	0.074	Fe^{2+}	0.074
Fe^{3+}	0.064	Co^{2+}	0.072
Ni^+	0.069	Cu^{2+}	0.096
Zn^{2+}	0.074	Ag^+	0.126
Sn^{4+}	0.071	I^-	0.220
Cs^+	0.167	W^{4+}	0.070
Au^+	0.137	Hg^{2+}	0.110
Pb^{2+}	0.120	U^{4+}	0.097

[*] Data from L. Pauling, *Nature of the Chemical Bond*, Cornell U. Press, 1945.

Note of Interest

Linus Carl Pauling was born in Portland, Oregon, on February 28, 1901. He published nearly 350 publications in the fields of experimental determination of the structure of crystals by the diffraction of X-rays and the interpretation of these structures in terms of the radii and other properties of atoms. He was awarded the Nobel Prize in Chemistry for his work on chemical bonding. He is generally regarded as the premier chemist of the twentieth century.

Problems

1. The boron ion, B^{3+}, has a radius of about 0.025 nm. Predict the coordination of B^{3+} in B_2O_3.
2. Predict the coordination in CsI.
3. Predict the coordination in AgI.
4. What coordination would be expected if $r/R < 0.1547$?

APPENDIX 10

Weibull Analysis

Weibull Analysis

For engineering use, it is important to determine not only the average strength of a material, but also the amount of its scatter. If the scatter is small, one can safely apply a stress only slightly below the average strength. On the other hand, if the scatter is large, the stress in service must be kept far below the average strength.

Weibull* suggested that in a large number of samples, the fracture data could be described by

$$P_s = \exp[-(\sigma/\sigma_o)^m], \qquad\qquad \text{A10.1}$$

where P_s is the probability that a given sample will survive a stress of σ without failing. The terms σ_o and m are constants characteristic of the material. The constant σ_o is the stress level at which the survival probability is $1/e = 0.368$ or 36.8%. A large value of m in equation A10.1 indicates very little scatter and conversely a low value of m corresponds to a large amount of scatter. This is shown in Figure A10.1.

Since

$$\ln(-\ln P_s) = m\ln(\sigma/\sigma_o), \qquad\qquad \text{A10.2}$$

a plot of P_s on a log(log) scale vs. σ/σ_o on a log scale is a straight line with the slope m as shown in Figure A10.2.

Example Problem 1:

Analysis of 100 tests on a ceramic indicated that 11 specimens failed at 85 MPa or less, and that 53 failed at a stress of 100 MPa or less. Estimate the stress that would cause one failure in 10,000.

* W. Weibull, *J. Appl. Mech.*, v. 18, 1951, p. 293 and *J. Mech. Phys. Solids*, v. 8, 1960, p. 100.

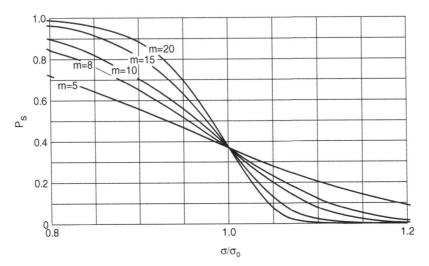

Figure A10.1. Probability of survival at a stress, σ, as a function of σ/σ_o. The relative scatter depends on the "modulus," m. From W. F. Hosford, *Mechanical Behavior of Materials*, Cambridge, 2005.

Solution: $P_s = \exp[-(\sigma/\sigma_o)^m]$, $\ln Ps = -(\sigma/\sigma_o)^m$, $\ln P_{s2}/\ln P_{s1} = (\sigma_2/\sigma_1)^m$, $m = \ln[\ln P_{s2}/\ln P_{s1}]/\ln(\sigma_2/\sigma_1)$. Substituting $P_{s2} = 1 - 0.53 = 0.47$ at $\sigma_2 = 100$ MPa and $P_{s1} = 1 - 0.11 = 0.89$ at $\sigma 1 = 85$MPa, $m = \ln[\ln(.47)/\ln(.89)]/\ln(100/85) = 11.50$.

$$(\sigma_3/\sigma_1) = (\ln P_{s3}/\ln P_{s1})^{1/m} = [\ln(0.9999)/\ln(0.47)]^{1/11.5} = 0.46. \, \sigma_3 = (46)(85)$$
$$= 39 \text{ MPa}.$$

The amount of scatter in fracture strength for brittle materials is much larger than for ductile materials. This is because the fracture strength, σ_f, of a brittle material is proportional to $KIc/\sqrt{\pi a}$, where a is the length of a pre-existing crack. The scatter

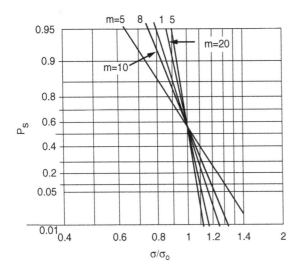

Figure A10.2. Probability of survival at a stress, σ, as a function of σ/σ_o. Note that this is the same as Figure A10.1, except here P_s is plotted on a log(log) scale and $\sigma/\sigma o$ on a log scale. The slopes of the curves are the values of m. From W. F. Hosford, *ibid*.

of fracture strength data is caused mainly by statistical variations in the length of pre-existing cracks in specimens.

The Weibull analysis can be used for analyses other than strength. It has been used in predicting the longevity of devices and financial analyses.

Note of Interest

Waloddi Weibull (1887–1979) was a Swedish engineer. In 1951 he presented his most famous paper on the Weibull distribution to the ASME, and in 1961 he published *Fatigue Testing and Analysis of Result*. He was awarded the Swedish Great Gold Medal as well as gold medals by the ASME.

Problem

1. Twenty specimens were tested to fracture. The measured fracture loads in newtons were:

   248, 195, 246, 302, 255, 262, 164, 242, 197, 224, 255, 248, 213, 172, 179, 143, 206, 233, 246, 295.

   a. Determine the Weibull modulus m and σ_0.
   b. Find the load for which the probability of survival is 99%.

Index

Printed in the United States
By Bookmasters